CIVIL

PE Exam

All-in-One Study Guide

2022 - UPDATED FOR COMPUTER BASED TESTING (CBT)

By: Juliene Sinclair, PE

To get a study schedule (and more!) straight to your inbox, register here:

www.peexampractice.com

❤ *DEDICATED TO MY MOM - FOR ALL THE THINGS* ❤

Table of Contents

An Intro You Want to Read

(Update: It's early 2022 and I am updating this book for the new computer based testing (CBT) exam style that you'll be experiencing. Although I passed the PE under paper and pencil style testing, and wrote this book accordingly, I have updated this study guide for you reflecting my research and understanding of the new CBT exam.)

I wrote this Guide to help you maximize your chances of passing the PE exam the first time. Before we get started, let me share my account of how I went from epicly failing the PE exam with a mind-numbing 30% to passing on the second try, when statistically very few pass in two attempts.

The first time - even after months and months of studying - I scored a percentage so low that, if I improved steadily, I would pass on the eighth time. I remember in college listening to a presentation wherein a successful PE, in an attempt to encourage us, regaled us with stories of repeatedly failing the PE exam and finally passing only on the eighth try. I give her kudos for persevering, but I remember thinking "No way! I'm going to pass that thing on the first try!". But once having failed, the fear that I too would be an eighth-timer crept over me.

The second time I took the exam, I passed with what felt like ease and grace. The NCEES second-time average pass rate was around only 30% when I re-took it. Clearly, this 30% knew something many of us didn't, and I wanted to learn the secret. The second time, I totally changed the *way* I studied - studying much less material but more thoroughly, and in a more relaxed and organized fashion. My goal was to study with calm and joy, feel awesome during the exam, and leave knowing I had passed. Success! The point of this? I am telling you that the second time around, I came up with a study method that took me from a 30% fail to passing, which prompted me to write this Guide for you.

There are two disclaimers due here. First, do not be shocked to find that my method is not shocking. The problem, I found, is that there is an overwhelming amount of material out there and so much conflicting advice, that the first time around I was uncertain how to proceed and simply dove in head-first trying to conquer it all. That may work for the few truly brilliant minds, but it did not for me. I was one of those people that did not really think I wanted the PE license, but deep down I now know I mostly just didn't think that I could pass the exam. I did pass the exam though. So whatever your story is, or however little you believe you can or will pass, I'm here to tell you that you can.

The second disclaimer is that I took a review course the second time around. A review course can keep you on schedule, and provide camaraderie and an instructor to ask questions. For the most part, however, I asked Google and Youtube questions because there wasn't enough time during class to ask all my questions. Outside of the actual material review, I didn't follow much of the instructor's study or exam advice because I became more confident in my own system that I was developing. I don't believe you need a review course, although I would never discourage anyone from taking one! I believe that if you narrow down *what* to study, *how* to study, and how to *organize* it, that anyone can pass with ease. This Guide was written to help you do just that.

Of course, YOU HAVE TO PUT IN THE WORK! This is no free ride. You're not going to just "figure it out" during the exam because that worked in college. You can't count on just getting lucky with the questions they ask or simply guessing the right answers when you don't know how to solve a problem. Although studying for the PE exam is a serious pain, you want to get it done and over with on your first attempt.

So, to focus you, the object of this Guide is to share with you the study habits that allowed me to go from failing the PE exam the first time to passing with a feeling of ease the second time. I hope that by knowing what I did to pass the second time, you can pass on your first!

This Guide is divided into two sections. The first section shares with you basic study issues and how to manage them - including general study guidelines, how to handle material overload, how to prepare your study material, the value of practice exams, time-wasting traps to avoid, and how to guess those last few head-scratcher problems on the exam. The second section contains three practice exams with solutions that include tips and tricks to help you navigate the material in the NCEES Handbook.

1. General Study Guidelines

This section discusses some big picture issues you should carefully consider before embarking on your study marathon.

1.1 Computer Based Testing (CBT)

Let's face it, change can be hard! Although at first it may be hard to see the benefits of CBT, I want to provide some relief for you and tell you a few of the big advantages of the new system:

- Year-round testing! No more waiting around for April or October, or having to make a choice between taking the exam and going to a friend's wedding.
- Individual testing environment. Just you and your cubicle. No more sharing long tables that get knocked by your neighbor's knees or hearing them open up that loud snack bar wrapper. With more testing centers and year-round testing, there will be far fewer people in the exam room with you and therefore far less pressure and built up tension.
- Faster reference searching. No more going to the Index and flipping back and forth until you find the right spot. Just Ctrl+F and off you go!

- Faster scoring. No more waiting 8+ weeks to get your results - instead you should be able to get your results within a couple weeks. Those eight weeks of limbo were painful for me, so this is huge!

On the flip side, here are some disadvantages:

- No additional reference material allowed. At first, this one made my heart sink for all future examinees mostly because I attribute my success to having created stellar customized reference material that I used while studying AND for the exam. There's definitely something about having actual paper in front of you instead of a screen. However, as I went through the process of updating this book for CBT, I realized that NCEES will likely be updating the exam accordingly by actually giving you more information in the problem statement. I say this because the NCEES PE Civil Reference Handbook (simply referred to as the "Handbook" from here on out) that will be provided to you in the exam is quite sparse on in-depth information.
- No customizing your own material. These first two disadvantages go together in the sense that if no material is allowed in, you don't have to worry about customizing your own material.
- No multiple choice. The solutions on the CBT test won't be all multiple choice, but rather a variety of answering methods like fill-in-the-blank, choosing a location on a figure, etc. This means that if you have to guess, the likelihood that you randomly guess correctly goes down as you won't have pre-provided solutions to guess from.
- The Handbook that is available to you in the exam contains important figures, tables, and equations but doesn't have in-depth explanations on subjects. Therefore, you'll have to go into the exam with a solid understanding of the main topics, which means you'll need to get intimate with the Handbook during your studies.
- More screen time. We all know screens are hard on your eyes and mind. This means that the exam day will have an extra exhaustion factor (maybe offset by the reduced examinees in room and more flexible scheduling?). You can build up your screen stamina during your studies by using a digital version of the Handbook.
- Computer glitches. No explanation needed.

If you study right, the pros will outweigh the cons!

1.2 Required Key Concepts

NCEES lays out for you the topics that are covered by the exam, so you shouldn't be wondering what you need to study. Start by going to the NCEES PE exam information website at: https://ncees.org/engineering/pe/. Select your discipline, then click on the applicable exam specifications.

This is where you start and what you're going to study, regardless of your affinity or distaste for a particular topic. You might say "I don't know anything about structural, so I'll just guess on those and aim for 100% on everything else!". This is a terrible idea. Of course we don't love many of the topics, but under testing pressure, it's extremely possible (and likely!) you'll even have difficulty with problems you know how to do. EVERY POINT COUNTS! Try your very best to know how to do everything on the NCEES list. You don't want to aim for 70%, you want to aim for 90%-100% and leave the exam room knowing you passed.

1.3 How-to-Study Basics

This section covers some study basics that aim to increase the *quality* of your study time.

1. *Struggle log*. Keep a running "struggle log" of problematic material - specific problems that don't quite make sense to you, units or concepts that you keep forgetting, *etc*. You know exactly what happens when you don't write something down because you *believe* you'll remember! Essentially, write down anything and everything with which you find yourself struggling during your studies, so you can return to them later and nail them down before test time. When you find yourself struggling with something, don't dwell on it and let it get you off track. Write it on your struggle log, stay on schedule, and return to these items later. Whittle away at this list and have it down to zero by exam day.

2. *Test-taking mindset*. Make your study time quality! This may seem obvious, but don't do anything during studying that you can not, or will not, do during the exam - like drinking alcohol, taking cigarette breaks, or even drinking too much caffeine. Try your absolute best to practice being in the test-taking mindset while studying. On this note, do your best to make your big study sessions outside of the comfort of your own home. The exam room doesn't have your dog, your spouse to talk to, or your snack-filled fridge. However, to make the study marathon bearable, do allow yourself some relaxing study sessions if it suits you, like leisurely following along with Youtube lessons while enjoying your morning coffee. And please, make time to do the things you enjoy to maintain a positive mental attitude.

3. *Choice of practice problems*. You'll notice when selecting study problems that you have a choice: you either can do really long, hard one-hour problems, for example as are found in the *PE Civil Practice Problems* by Michael Lineburg; or, you can do six-minute problems (the average amount of time allotted per exam problem). I kept generally with the latter so I could cover more problems and also because these types of problems are more similar to those on the exam. However, know the value of the one-hour problems: they often fully cover a concept with the multiple answer parts, a, b, c, *etc.* For example, a one-hour problem may have one part about void ratio, the next part about porosity, the next part about saturation, *etc*. By its nature, the exam can only ask one question (no multiple part questions), but you'll never know which part of what topic they'll ask, which is where these one-hour problems come in handy.

4. *Make a choice.* You'll hear so many different strategies on studying (including my own), but in the end you simply must pick an approach then try not to learn too much or too little. If you're learning too much, you're wasting time because much of it may be outside the scope of the exam. If you're learning too little, then you're either overwhelmed or just procrastinating. If you understand the concepts and can handle the material in the specifications *under testing conditions*, you are learning the right amount.

Don't get bogged down calculating different approaches to studying the material. Make a decision early on about how much time you'll spend on depth material and how much on breadth material - then keep to your schedule. My first time, I had been recommended to focus on the depth, because much of that depth is also in the breadth. This didn't work for me the first time because I let much of the other material fall by the wayside. When I took the exam the second time, my review course allotted two months for breadth and one month for depth, so that's what I did. I felt like one month for depth was a bit hectic and too short, but it worked out as we had already covered much of the material in the breadth section.

1.4 Quality vs Quantity

The general advice is to study 300 hours for the exam. This breaks down to approximately 3.33 hours per day for three months. Ugh, that's a lot of hours! This could easily lead you to believe that quantity of studying is more important than quality. The first time, I probably studied about 300 hours, but I was overwhelmed by the amount of material that I thought I had to know and thus invested time in quantity instead of quality. I studied all the problems under the sun and got off track into the weeds studying for those "curveballs" not related to the core material, but which I feared would be on the exam. This method caused me to fail the first time because I knew a lot of things, but was not fully confident with the core concepts.

Here's an important tip: don't spend much, if any, time on the curveballs during your study time, otherwise you'll waste precious time understanding critical topics like Bernoulli's equation or soil classification. If you study right, you'll be going through the Handbook enough to know how to locate and solve these curveball problems using keywords and common sense. The winning strategy for handling these curveballs in the exam is simple: find the keyword in the problem and search it in the Handbook, and if you don't find it there, make the best guess you can and move on. Some curveballs are actually just testing your common sense, so if you don't see a keyword, just think the problem through and choose the most sensible answer. Then, move on so you can spend your time on the problems you know you can get right.

Remember, you need to spend your study time hammering problems that go over the *key* concepts outlined by NCEES. Know these key concepts inside and out, how to identify different variations, and how to solve them in under six minutes. Essentially, be able to do them in your sleep!

2. About Practice Exams

The first time I studied, I don't think I did one practice exam under fake pressure and with a time limit. I just moseyed my way through them and called it good. I had never failed a test before, so I was confident that I would figure it out like I always did. This proved to be a fatal mistake for so many reasons. Because I hadn't simulated testing conditions and hadn't become intimately familiar with the key material, my stress overtook my senses during the exam and I found myself fumbling through my reference material and doubting myself every step of the way. Most importantly, I couldn't get on autopilot because I had never practiced getting on autopilot during my studies. Finally, I had not prepared for potential time wasters, which I discuss below.

The second time I studied - against the recommendation of my review course instructor - I took practice exams until I was blue in the face. About a month into studying the breadth, I took a practice exam. I was super serious about it. I timed myself and straight up faked that I was in the actual exam. My goal for the second time around was to stay ultra calm and focused during the exam and leave the test knowing I had passed. The only sure way to do this was to take a zillion practice exams and pass them, so when the real one came, it was just another practice exam … kind of.

In addition to preparing you for stress conditions, simulated practice exams quickly allow you to see where your weaknesses lie. I re-took the same practice exams over and over, not only until I passed, but until I achieved 90-100%. It may seem like this time would be better spent exploring new material, but there is just so much material and such a time crunch that even when you are seeing the exact same problems, you still make the same mistakes, forget the same stuff, and fall into the same traps. By re-taking the same practice exams until you ace them, you are hammering into your brain the key concepts while also learning to mitigate time wasters and panic makers.

I started taking breadth portion exams a month into studying, and took them about once a week throughout the remainder of my studies. Of course, I hadn't covered all the material by the time I started taking them, but this got me ahead of the curve because I had to become at least familiar with material I'd be seeing soon enough. Once I started studying the depth, I did the same thing with the depth practice exams. Then only in the final month, after doing a couple of depth practice exams, did I take full-length practice exams. Not only was this to practice mental endurance, but also to stay fresh on the breadth material. Someone in my review class said that they didn't even remember the material they studied one month ago, which I found surprising because all of the practice exams were keeping me fresh!

Speaking of mental endurance, sitting down for eight hours a day at work is one thing, but sitting down for eight hours under immense exam pressure can wear any soul down! Practice exams are crucial to developing the mental fortitude to get you through the test. Of course, being under the pressure of the real exam will kick you into gear and the adrenaline will carry you to some extent, but if you don't have everything else on autopilot, the adrenaline can only make things worse. Remember, the goal is to feel confident in the exam and walk out knowing you passed!

Not only are actual practice exams useful, but you can even turn homework problems into practice exams. For example, do a homework problem at a leisurely pace, and then do it in six minutes. Don't have four hours after work to take a practice exam? Do ten problems in an hour. Do five problems in one-half an hour. Whatever time you have, do that. It is amazing what happens when you put yourself under time pressure. This time pressure forces your brain to go into efficiency mode, and it's in this mode that time-wasters and your problem solving process become clear.

Don't just take it from me, though. Research has actually shown that practice testing is the most effective way to improve a test score!

2.1 CBT Practice Tests

One of the downsides to the new CBT testing method is that most, if not all, of the practice exams out there (like the ones in this book) are pencil and paper. The CBT exam will certainly be a bit different than pencil and paper, but the pencil and paper practice exams are still crucial to utilize as it is not so much about the format of the exam as it is your understanding of the topics. It is highly recommended that you have the Handbook open on your computer/laptop and use it while you take your pencil and paper practice exams. This will help you get used to where different topics, equations, figures, and tables are located in the Handbook. However, this is going to require you to be diligent about not getting distracted by any other tabs or programs you have running on your computer!

3. Time Wasting Traps

While you study, you want to reign in the things that waste your time, which is different for each person. I will share a few of my own time-wasters and the rules I put in place to reign them in. Before we begin though, I want you to note two points: 1) some of these rules may sound like time-wasters in and of themselves, but if it takes a few moments longer to prevent a mistake, consider it a win; and 2) your own specific time-wasters will only become abundantly clear under time pressure, so this is an additional reason to focus on taking practice exam under exam conditions. Here are a few ways to be ultra time-efficient:

1) <u>Read the problem:</u>
   This is an obvious one, but when you're under exam pressure it's easy to move *too* quickly and just skim the problem. Not only do you want to carefully read each question all the way through, but it's also extremely helpful to underline important information that will clarify the problem. Underline the units, parameters, and the actual question. Make sure the methodology you are about to employ is the correct one before you start calculating.

2) <u>Do easy problems first:</u>
   Go through the *entire* exam and knock out the easy ones as you go. This boosts confidence and warms up your brain. Then, keep going back through the exam doing the next easiest problems.

3) <u>Do not get stuck on one problem:</u>
   If you're getting stuck on solving a problem, stay calm and move on. Some questions you know you can figure out, but they just need to incubate a bit - do these ones after the easy ones, but before the "I have no clue" ones. Usually another problem will knock something loose and you can easily go back and solve the problem that just needed a bit of mulling over.

4) <u>Focus on one problem at a time:</u>
   When you decide to move on from a problem you haven't yet solved, do not keep going back to it because you think you may have figured it out. You have many other problems to solve, so don't get sloppy. Focus on one problem at a time.

5) <u>Double check your answers:</u>
   Check your answers before moving on to the next question. If the second result doesn't match the first, calculate it again. This may sound like it is a time-waster in and of itself, but if you want to leave the exam knowing you passed - then checking your answers will help you feel confident that you did. If you write neatly and completely, including units, then checking your answers only takes a few seconds.

6) <u>Pay attention to units:</u>
As an engineer, you very well understand how important units are, but on the exam this importance is on hyper drive. When you read the question all the way through, look at the units and what conversions you'll have to make. If you get an answer that doesn't make sense with the answer options, sometimes it will be obvious that you didn't make a conversion. But beware - some of the wrong answer options may be based on typical errors in conversions. Also, under the exam pressure, it's easy to feel so rushed that you don't want to write down the units as you work - but writing the units down helps you see where you made a mistake and can save you tons of time in the end!

7) <u>Learn how to use your calculator:</u>
Learn how to use your calculator efficiently, which may mean reading the manual. Set it to decimal mode. Learn how to solve for x. Learn how to save numbers. You have enough to think about, so let your calculator do the heavy lifting when possible.

8) <u>Avoid panic makers:</u>
This was huge for me because I am a "bad test taker", which means that I easily forget things I know because I get so nervous. During your studies, make note of what causes you to panic (see the discussion about a "struggle log" in the 1.3 How to Study Basics section). For example, maybe you mix up gravity or the density of water in metric units; maybe you mix up unit weight and density; or, maybe you for some reason always struggle to find the right equation for a certain problem. Whatever it is, panic will bring you down during the exam so make sure to eliminate these throughout your studies by practice, practice, practice!

9) <u>Prevent simple mistakes:</u>
There are plenty of opportunities to make simple mistakes with all this material, and also plenty of opportunities to prevent them. Never do things in your head that could be done on paper or with your calculator, including the simplest of things like the ol' two place decimal bounce when going from ft/ft to slope percentage, for example.

10) <u>Rounding:</u>
Don't worry about rounding until the very end of your calculations. Store original numbers in your calculator and use them throughout your calculations - and then round at the very end. The solutions in this Guide have rounded calculations for simplicity. However, the solutions on the exam should be far enough apart to account for slight differences in rounding. It is generally safe to store original numbers in your calculator throughout your calculations and then round at the very end. Also, use π and *e etc.* instead of their rounded counterparts.

4. Guessing and Bubbling

4.1 Guessing

Although the aim is to not have to guess, it's likely you won't have time or be stumped for a couple questions. You'll hear different theories about how to guess and, in the end, it really doesn't matter. However, you must have a strategy for guessing so when the time comes, you're not wondering *how* to guess. If you do a problem to the best of your ability and come up with an answer that isn't offered, then unfortunately you'll have to guess. Only guess answers at the very end of the exam when you know that you absolutely won't have time to finish calculating them. The reason for this is that sometimes something will get jolted awake in your brain half way through the exam and you'll be able to figure it out.

Hopefully this goes without saying - don't leave any problem unanswered!

5. How to Use This Book

I hope that this book helps you see through the overwhelming fog of studying for the notorious PE exam. This Guide is meant to help you focus on what and how to study, pitfalls to avoid, time saving tips, and loose but important ends like how to guess. Of course, take and leave any of the tips I provide here, and you'll even probably come up with your own!

The majority of the solutions to the practice exams in this book reference the NCEES Handbook, which is what you'll have available during the exam and must be your go-to reference. As was mentioned above though, the Handbook does not have in-depth and clarifying text and so it is recommended to study with a copy of the Civil Engineering Reference Manual (CERM) by Michael Lindeburg. The CERM was the go-to manual for the PE Exam when it was pencil and paper, and is a fantastic (although sometimes overwhelming because of its size) reference for you to learn from. The majority of the solutions in this Guide have a quick note referencing the corresponding CERM Chapter, which has been included with the assumption that many people will still be using the CERM to study. The solution notes reference the CERM 14th Edition. Based on our review, this is also appropriate for use with the 15th Edition. If you are using an earlier edition of the CERM, the practice exams in this book will still be helpful, but some of the equations may be slightly off.

The practice exams cover the majority of "must study" topics but also throw you a few curveballs so you have some practice with those, too. Many of the solutions include notes with some additional information for you that may be helpful for your studies.

6. Register Book to Receive a Study Schedule (and more!)

Go to www.peexampractice.com and follow the instructions to register this book, which will sign you up to receive additional material straight to your email inbox. Included in this information will be a suggested study schedule, which main topics to cover, which equations to study, and additional helpful resources. Although the exam is offered year-round now with the transition to CBT, we still recommend studying for about four months in advance of when you take the exam. To accommodate different study schedules, all of our additional resources will be provided to you at once. It will be up to YOU to stick with your schedule and stay on track with your studies!

All of the additional material we'll be sending you will reference the CERM, which might be frustrating for you considering that you won't be able to bring the CERM into the exam with you. The main reason for this is that, as was previously mentioned, the NCEES Handbook does not have text accompanying equations, figures, and tables, nor in-depth explanations of the various topics. In order to come up with the correct answers on the exam, it's important that you actually understand the material, which will help you properly come to the correct calculations. Therefore, it's highly recommended that you still get yourself a copy of the CERM and the Practice Problems for the Civil Engineering PE Exam: A Companion to the Civil Engineering Reference Manual. Although these books aren't cheap, they'll save you a whole bunch of headache trying to gather, collect, and understand study material from all the various sources that exist. You can always re-sell them after the exam; however, you may want to hold on to the CERM considering it'll definitely come in handy during your professional career!

No upsells or promos on the email list - we promise.

7. A Closing Note

I would be lying if I said that finding out I passed the PE exam, after epicly failing once, wasn't one of the happiest moments of my life. I had spent so much time and had poured so much of myself into studying the first time, that I was absolutely crushed to have gotten a score so low that I likely could have scored the same without studying at all. I hated that I had unknowingly squandered so much time and energy, and worried about what it meant about my ability to ever pass. To have gone from 30% failing to not only passing, but feeling super confident during the exam and walking out knowing I had passed, was a serious reason for me to celebrate. In my excitement of passing, I started writing everything down knowing that I wanted to write a book to help others pass as well. I truly hope you find this book helpful.

Remember, the point of the exam is to find the right answer as quickly as possible. What you're going to do is study your heart out, go and sit in a room, and take the test. You will be done in nine hours, including lunch. No, it's not going to be fun. Yes, it'll be worth it!

If this book helps you, please make the time to leave this book a review on Amazon! If you don't find this book helpful, please contact us through www.peexampractice.com and let us know where we can improve.

Finally, practice positive self-talk during your studies! You have everything within you and all the necessary resources available to you to pass on your first try.

YOU GOT THIS.

Errata: This book has been heavily reviewed by multiple professional engineers. We are very grateful to readers who notify us of any potential omissions and/or errors. Your feedback allows us to improve the quality of this book and thus your study time. Helping you pass is our priority! Contact us at www.peexampractice.com to report errors and/or omissions, request corrections and/or addendums, or provide suggestions on improvements. Thank you for your understanding and support!

1) The force (kN) acting in Member CD in the truss shown below (neglect self weight of truss) is most nearly:

A) 7.41 (compression)
B) 7.41 (tension)
C) 5.25 (compression)
D) 5.25 (tension)

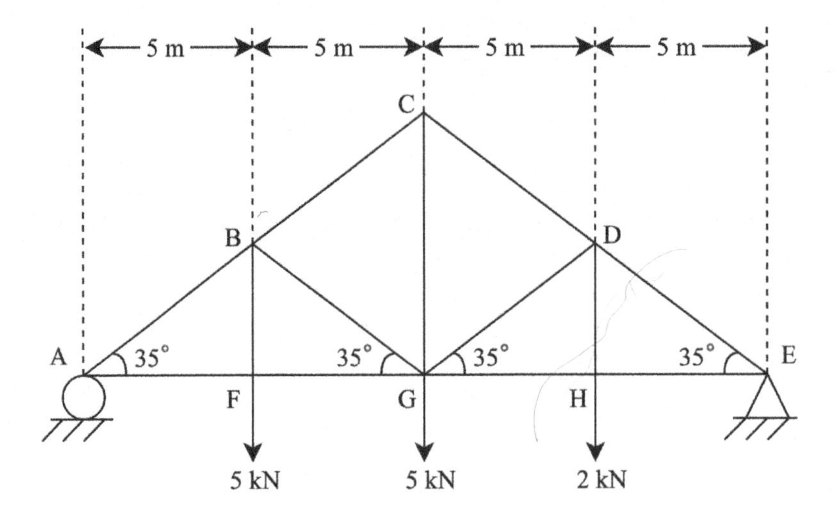

2) What is the most correct sequence for the construction of a wooden building?

A) I, IV, III, II, V
B) V, I, II, III, IV
C) II, I, III, IV, V
D) I, III, IV, II, V

I. Site clearing
II. Rough grading
III. Concrete foundation
IV. Excavation
V. Wood framing

3) The construction of a roadway embankment requires a compacted fill of 95% relative compaction and water content equal to the optimum moisture content, ±3%. The standard Proctor test curve is shown below. Most nearly, what is the relative compaction (%) in the field and is it acceptable?

A) 89; acceptable
B) 94; not acceptable
C) 94; acceptable
D) 99; acceptable

Water content of in-situ soil on site = 11%

ϱ_{dry} = 15.85 lbm/ft^3

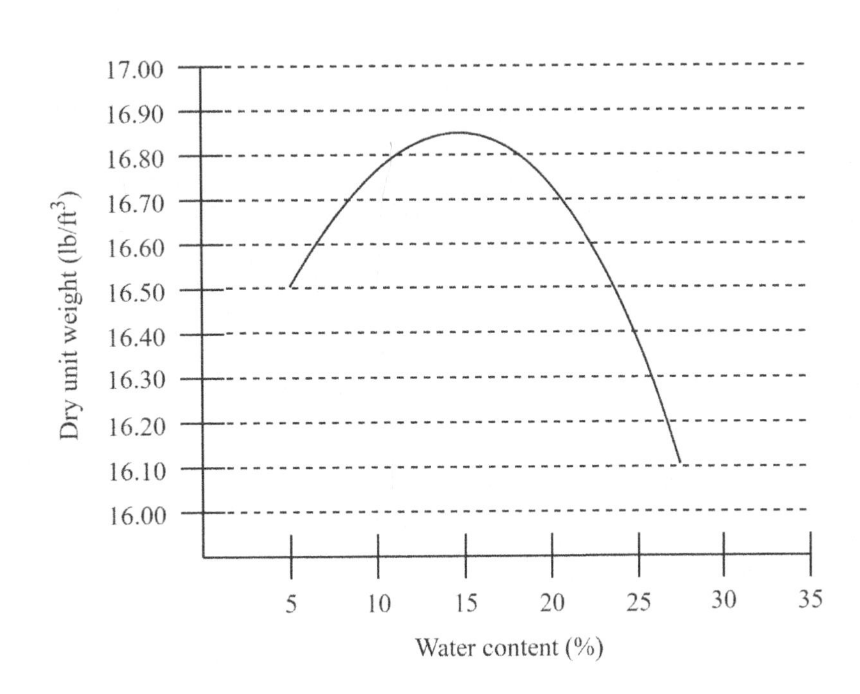

4) A 20 m long structural steel member is used as a slender column to support a load of 325 kN. There are no intermediate supports. One end is pinned, the other is fixed against rotation but free. The required moment of inertia (m^4) is most nearly:

A) 65.1 x 10^3
B) 5.27 x 10^{-4}
C) 13.0 x 10^{-5}
D) 13.2 x 10^{-5}

E = 20x10^{10} Pa
FS = 2

5) Using the Terzaghi-Meyerhof bearing capacity equation and the Terzaghi bearing capacity factors, what is most nearly the ultimate bearing capacity (kPa) of the continuous strip footing shown?

A) 951
B) 1,224
C) 1,646
D) 25,701

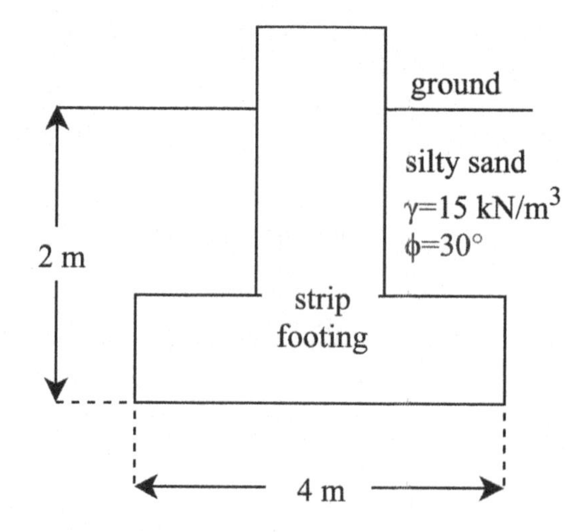

6) A horizontal curve for a highway has the elements shown below. The intersection angle (degrees) is most nearly:

A) 55.55
B) 27.78
C) 55.25
D) 56.00

M = 138.10
D = 4.78°
T = 631.35 ft

7) The effective stress (pcf) at the middle of Layer 3 is most nearly:

A) 2,058.80
B) 2,164.40
C) 2,682.80
D) 15,630.00

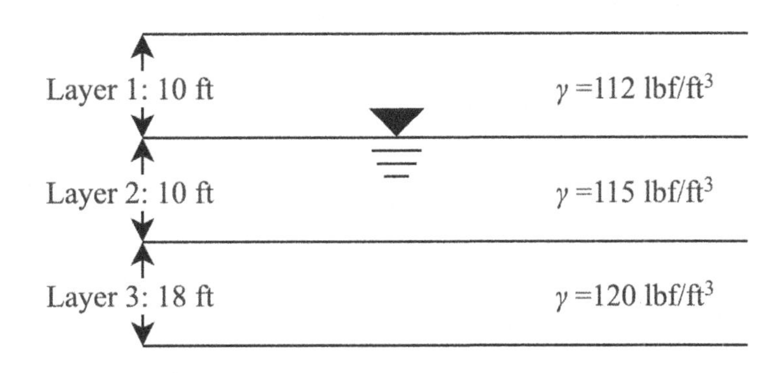

8) Which of the following most adequately describes the construction procedure for concrete using the following steps?

A) VI, V, I, VII, IV, III, II
B) VII, I, IV, V, II, VI, III
C) VII, II, III, VI, I, IV, V
D) VII, I, II, III, IV, V, VI

I) Prepare the formwork
II) Pour the concrete
III) Disassemble the formwork
IV) Place steel rebar
V) Prepare concrete mix
VI) Cure the concrete
VII) Preparing the site

9) The cantilever frame shown is supported by a 10m x 10m footing. Neglect the thickness and weight of the footing. The maximum and minimum tensile stress equations are given below. The maximum normal tensile stress (kN/m^2) below the footing is most nearly:

A) 0.204
B) 0.404
C) 1.084
D) 1.087

$$\sigma_{max,\ min} = \frac{F}{A} \pm \frac{Mc}{I_c}$$

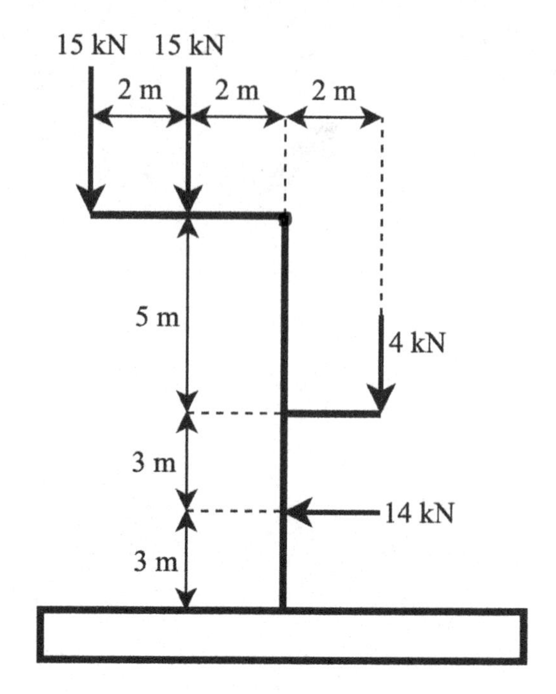

10) Water flows at 135 L/s through a schedule-40 steel pipe that gradually enlarges in diameter from point A to point B. All minor losses are insignificant. The total energy at point B (J/kg) is most nearly:

A) 264
B) 265
C) 291
D) 301

Point A
Elevation = 456 meters
$P_A = 70$ kPa
$E_{A,\,total} = 94.50$ J/kg

Point B
Elevation = 478 meters
$P_B = 49$ kPa
$D_B = 0.50$ meters

11) A slope is shown below with soil parameters. Most nearly, what is the Factor of Safety against shear failure using the Taylor method?

A) 0.95
B) 1.74
C) 1.76
D) 7.07

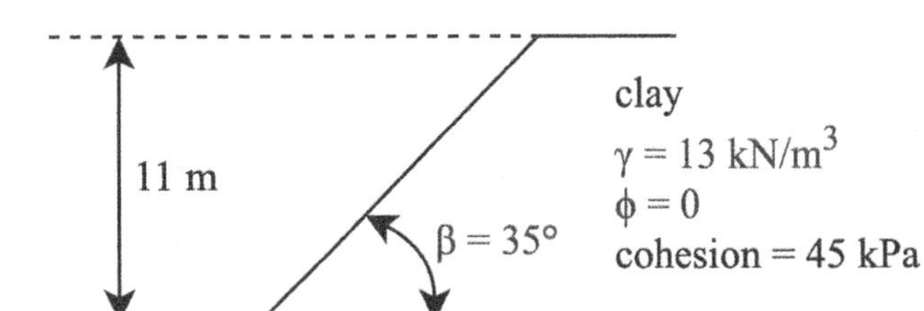

12) The following information is submitted for a proposed concrete mix. The absolute volume of fine aggregate (ft^3) in 1 yd^3 of fresh concrete is most nearly:

A) 3.40
B) 4.11
C) 9.52
D) 17.48

Cement	
Specific gravity	3.10
Cement content	7 sack mix
Fine aggregate	
Specific gravity	2.48
Fineness modulus	2.75
Absorption	3.5%
Coarse aggregate	
Volume	8.96 ft^3/yd^3
Specific gravity	2.70
Dry	1,418 lbf/yd^3
Water-Cement Ratio	0.39
Air content	3.75%

13) Sewer and water pipes are designed to have 18 in. of vertical separation (outside-of-pipe to outside-of-pipe). The sewer main has been installed per design with a flowline elevation of 4782.28 ft at a location where a water main crosses above the sewer. The water and sewer pipes both have an inner diameter of 12 in. and outer diameter of 13.75 in. The rod reads 10.76 ft on the top of the water main. Assuming a flat existing ground, the ground elevation (ft) directly at the top of the trench is most nearly:

A) 4782.28
B) 4786.00
C) 4793.26
D) 4796.76

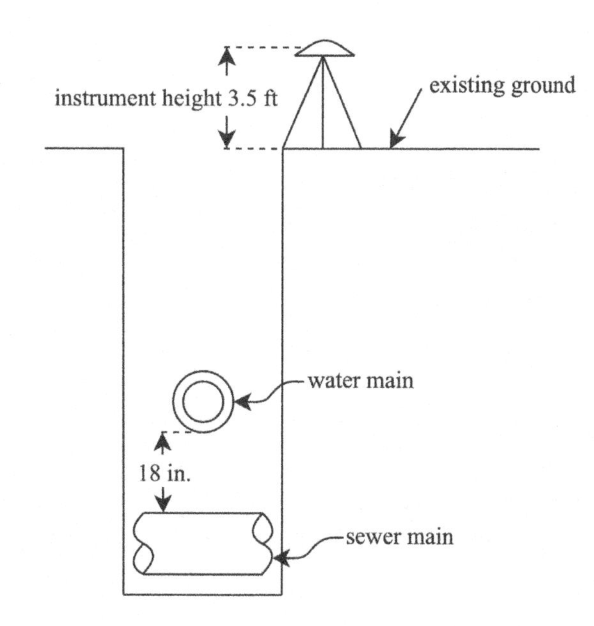

14) Settling is generally due to consolidation of the supporting soil. Which are the three distinct periods of consolidation?

A) Primary, secondary, and tertiary
B) Immediate, primary, and secondary
C) Immediate, overconsolidation, and normal
D) Vertical, horizontal, and secondary

15) An 8 ft long cantilever beam is loaded by 250 lbf at 2 ft from its free end. The cross section is 8 in. wide by 4 in. high. The maximum deflection of the beam is most nearly:

$E = 1.5 \times 10^6$ psi

A) 0.24
B) 0.49
C) 0.73
D) 42.7

16) A bridge is proposed to cross over the sag vertical curve shown below. If 22 ft of clearance is required, what is most nearly the lowest elevation (ft) that the bottom of the overhead tram bridge may have at station 22+20?

A) 698.35
B) 698.41
C) 712.43
D) 724.53

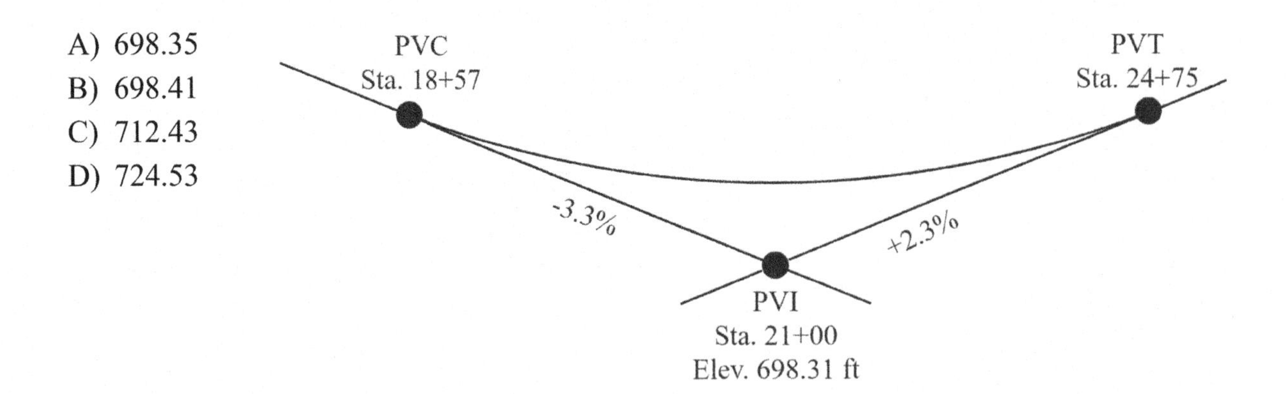

17) An 8.5 ft tall retaining wall to be made of normal weight concrete is being poured on a day with a measured temperature of 92° F. The concrete mixture is Type I with no added retarders. The concrete lateral pressure distribution equation is given below. The reduction in hydrostatic pressure (psf) due to pour and set up rates is most nearly:

A) 541
B) 600
C) 675
D) 3,000

$$p_{max,psf} = 600C_w \leq C_w C_c \left(150 + \frac{9000R_{ft/hr}}{T_{°F}}\right) \leq 150h_{ft}$$

$C_w = 1.0$
$C_c = 1.0$
$R = 4$ ft/hr
$\gamma = 150$ lbf/ft^3

18) A construction site near a waterway requires temporary best management practices (BMPs) for erosion and sediment. A slope drain is used to control the flow of water off of the construction site. Which of the following is most appropriate to control erosion at the end of the slope drain?

A) Sediment structure
B) Temporary berm
C) Silt fence
D) Temporary seeding

19) $550 is compounded monthly at a 5% nominal annual interest rate. Most nearly how much ($) will have accumulated in 4 years?

A) 450
B) 669
C) 672
D) 2,875

20) Primary consolidation occurs in clayey soils due primarily to:

A) Extrusion of water from the void spaces
B) Plastic readjustment of the soil grains
C) Compression of the soil grains
D) Specific gravity of the soil

21) The following volumes shown are the cross sectional areas at the respective stations along a roadway. Most nearly, what is the total volume excavated (yd^3)?

A) 4,900
B) 4,919
C) 132,795
D) 14,755

Station	Area (sf)
12+50	0
12+96	510
13+25	234
15+46	528
15+55	498
16+01	435

22) A frictionless retaining wall is shown. The resultant of the active earth pressure (kN) acting on the wall is most nearly:

A) 37
B) 99
C) 101
D) 168

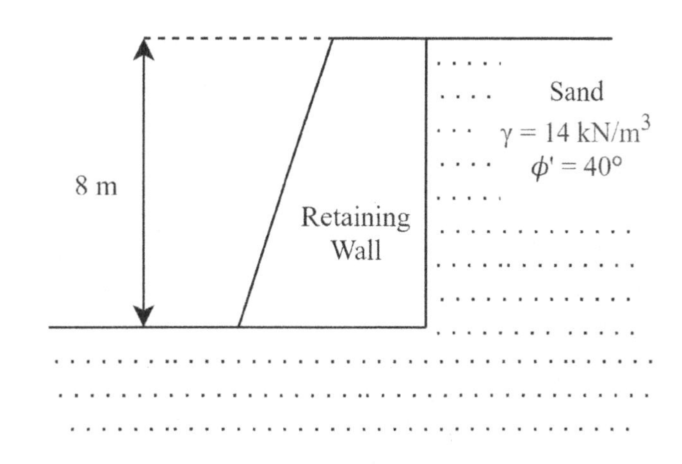

23) Which following statement regarding transportation Levels of Service is most accurate?

A) Level A represents severely impeded flow
B) Level B generally has the maximum capacity
C) Economic considerations favor less obstructed levels of service
D) Political considerations favor less obstructed levels of service

24) The short column shown carries a maximum concentric load of 200 kN and has a square cross section. The allowable strength of the column material is 200 kN/m^2. The minimum required side length (m) of the square is most nearly:

A) 1.0
B) 1.10
C) 2.0
D) 2.10

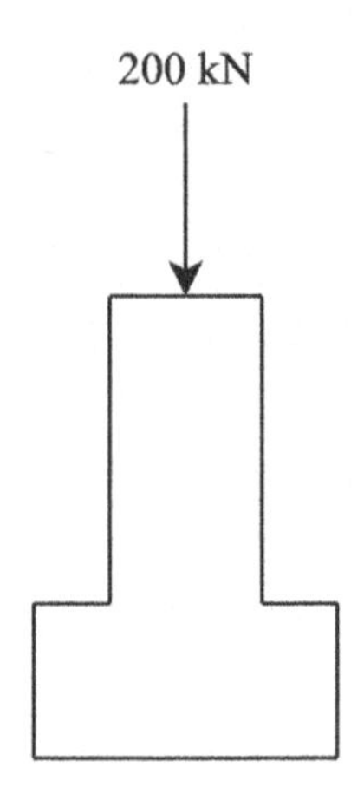

25) Which statement is false?

A) The Bernoulli equation states that the sum of the pressure, velocity, and elevation heads is constant.
B) The *hydraulic grade line* depicts potential energy (position plus pressure head) at all stations along the system.
C) The *velocity head* is equal to the elevation of the *energy grade line* minus the elevation of the *hydraulic grade line.*
D) In the Bernoulli Equation, the difference between the hydraulic grade line and the energy line is the P/γ term.

26) The mechanical properties of structural steel are generally determined by what kind of tests?

A) Tension
B) Compression
C) Torsion
D) Heat

27) Which type of reinforcement is appropriate for concrete that is exposed to weather or in contact with the ground?

A) No. 14 and No. 18 bars
B) No. 6 through No. 18 bars
C) No. 11 bar and smaller
D) Stirrups, ties, and spirals

28) Soil extracted from a site at a depth of 15 m has an Atterberg liquid limit of 45.2 and plastic limit of 14.9. The grain size distribution is shown below. What is the classification of the soil according to the AASHTO system?

A) A-7-6
B) A-6
C) A-2-5
D) A-8

US sieve size	**Cumulative percentage retained**
No. 4	0%
No. 10	7.1%
No. 40	18%
No. 200	36.7%

29) A steel layout plan is shown including the section lengths and weights. The total weight (lbs) of the steel is most nearly:

A) 33,145
B) 38,160
C) 51,475
D) 52,225

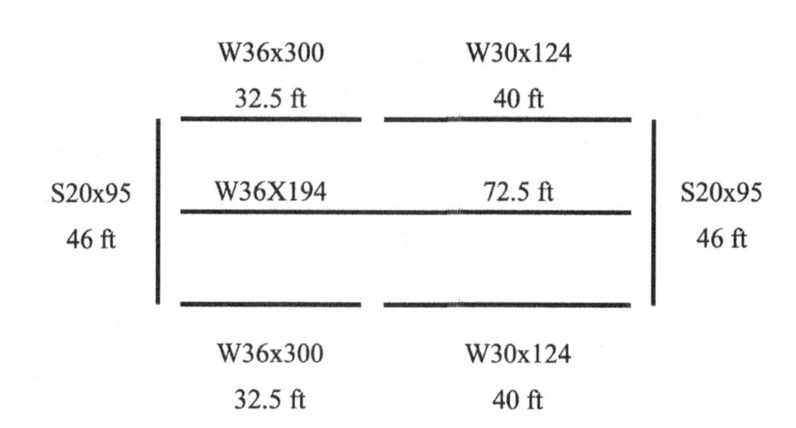

30) An 85-ac watershed has two distinct soil types shown below. The storage capacity of the soil (in.) is most nearly:

A) 0.90
B) 1.30
C) 3.80
D) 11.30

24 acres	61 acres
Straight row broadcast legumes Poor hydrologic condition	Straight row small grain Good hydrologic condition
I = 0.35 in/hr	I = 0.16 in/hr

31) Which is the best rebar reinforcement for the retaining wall shown below?

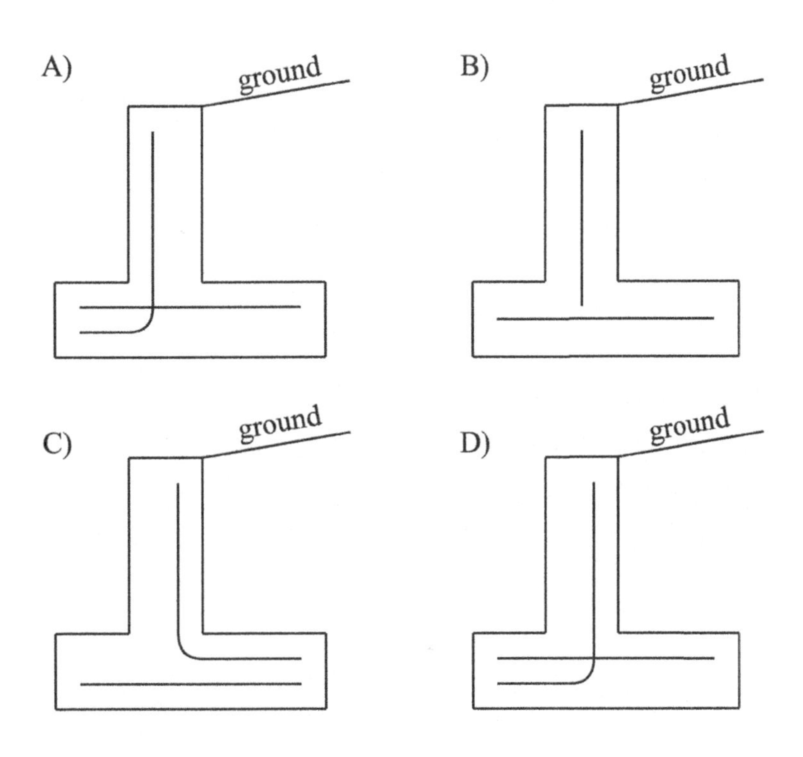

32) The void ratio of the moist, in situ soil sample with the following characteristics is most nearly:

A) 0.31
B) 0.44
C) 0.94
D) 1.96

Mass:	58.72 lb (as sampled)
	49.63 lb (after oven drying)
Volume:	0.60 ft^3 (as sampled)
Specific gravity of solids:	2.59

33) Which of the following statements is false regarding activity-on-node networks?

A) A critical activity is one without zero total float.
B) Delaying the starting time of an activity on the critical path will delay the entire project.
C) Float time is the amount of time that an activity can be delayed without affecting the overall schedule.
D) Float time is defined as float = late start - early start

34) It is required that 10 ft of separation be maintained between the edge of the house foundation and the top of the trench slope. If the underlying soil has the type and depths shown, the distance (ft) between the house foundation and bottom of the 6.5 ft deep trench is most nearly:

A) 12
B) 14
C) 15
D) 18

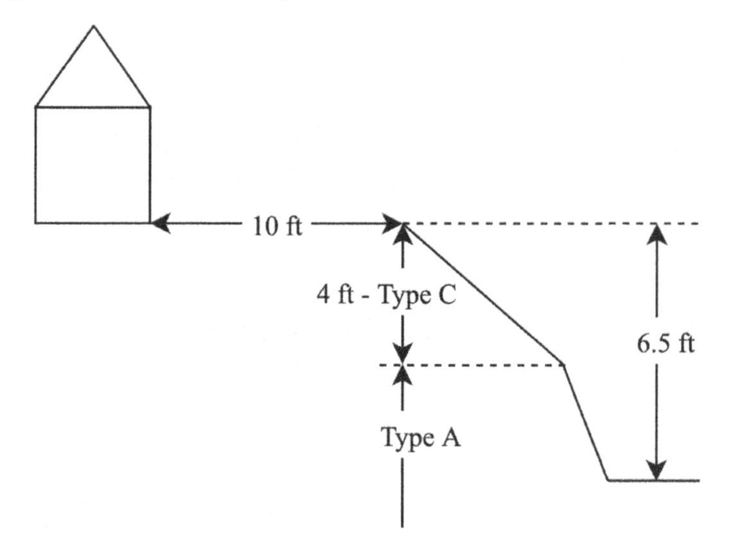

35) Which of the following is used in determining the experience modification rate?

A) Secondary losses
B) Unpaid losses
C) Reserved losses
D) Unexpected losses

36) The probability of flooding is 1% in any given year. The standard tenure of a facility manager is 30 years. During the 60-year design lifetime of a facility, the probability of flooding is most nearly:

A) 0.40
B) 0.45
C) 0.55
D) 0.56

37) Most nearly, what is the entrance velocity (ft/s) of a corrugated metal culvert with no headwall and an entrance head loss of 9 ft?

A) 12
B) 20
C) 25
D) 89

38) A retention pond is sized to hold 1,600 gallons of water. If a storm produces a runoff flow of 0.56 cfs that is directed into the pond, most nearly how many minutes will the pond take to fill?

A) 0.2
B) 6
C) 8
D) 119

39) Most nearly, what is the shear force (kN) at support A for the simply supported beam shown?

A) 66
B) 88
C) 89
D) 99

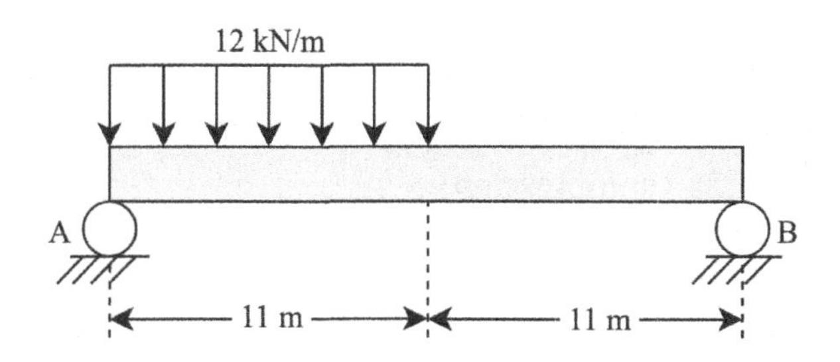

40) A sewer pipe line has a diameter of 15 in., a slope of 0.40%, and a Manning roughness coefficient of 0.016. The uniform flow inside the pipe has a depth of 3 in. The average velocity of the flow inside the pipe (ft/s) is most nearly:

A) 1.66
B) 2.34
C) 5.28
D) 16.68

- END OF EXAM VERSION 1 -

1) The force (kN) acting in Member CD in the truss shown below (neglect self weight of truss) is most nearly:

A) <u>**7.41 (compression)**</u>
B) 7.41 (tension)
C) 5.25 (compression)
D) 5.25 (tension)

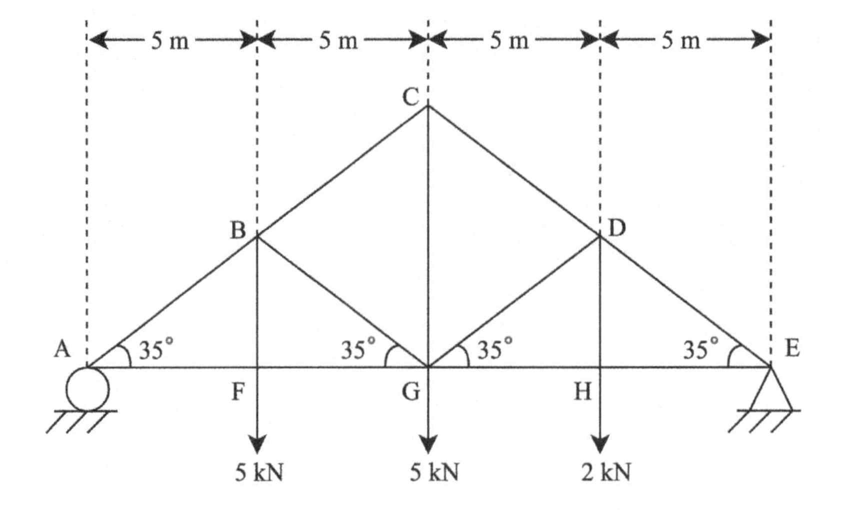

FIND: Force acting in Member CD (kN)

Step 1) Search for and go to the *Method of Sections* in the Handbook.

Step 2) Using the method of sections, make a cut through DC, DG, and GH:

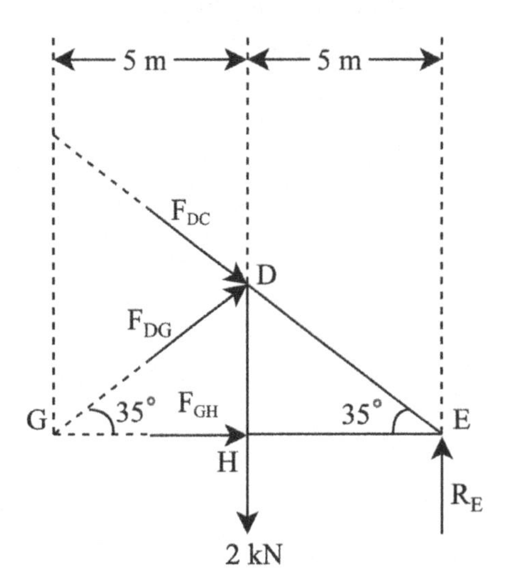

Step 3) Solve for R_E by calculating the moment around support A:

$$M_A = \Sigma Fd = 0$$

$$20\,m\,(R_E) = 5\,m\,(5\,kN) + 10\,m\,(5\,kN) + 15\,m\,(2\,kN)$$

$$R_E = 5.25\,kN$$

Step 4) Solve for F_{DC} by calculating the moment around support G:

$$\Sigma M_G = \Sigma Fd = 0$$

$$10\,m\,(F_{DC} sin35\,\circ) + 5\,m\,(2\,kN) = 10\,m\,(5.25\,kN)$$

$$F_{DC} = 7.41\,kN\text{(compression)}$$

Notes:

- CERM Chapter 41
- The chance of getting a truss problem on the exam is high!
- In common trusses, the truss bottom chord is in tension and the top chord is in compression. Therefore, answers B and D can be eliminated automatically.
- Remember it is expected to start off assuming the direction of the forces. A member is in compression when a member force points towards the joint it is attached to. The member is in tension if the force points away from the joint it is attached to. If your calculated magnitude is negative, it is the opposite of what you assumed.

2) What is the most correct sequence for the construction of a wooden building?

A) <u>I, IV, III, II, V</u>
B) V, I, II, III, IV
C) II, I, III, IV, V
D) I, III, IV, II, V

I. Site clearing
II. Rough grading
III. Concrete foundation
IV. Excavation
V. Wood framing

Notes:

- Even if you're not familiar with building construction, this type of problem can be easily thought through.

3) The construction of a roadway embankment requires a compacted fill of 95% relative compaction and water content equal to the optimum moisture content, ±3%. The standard Proctor test curve is shown below. Most nearly, what is the relative compaction (%) in the field and is it acceptable?

A) 89; acceptable
B) <u>94; not acceptable</u>
C) 94; acceptable
D) 99; acceptable

Water content of in-situ soil on site = 11%

$\gamma_{dry} = 15.85 \text{ lb/ft}^3$

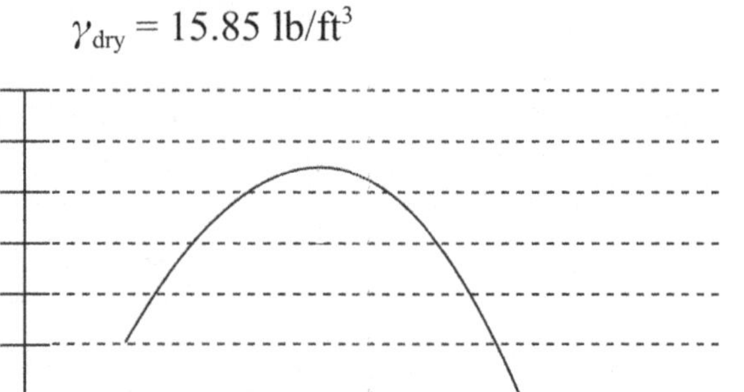

FIND: Relative compaction, RC (%)

Step 1) Search for *relative compaction* (RC) in the Handbook and navigate to *Chapter 2.9.2 Field Compaction, Field Density Testing.* The first equation is for RC:

$$RC = \frac{\gamma_{d\,field}}{\gamma_{d\,max}} \cdot 100$$

Step 2) Determine that maximum unit weight is approximately 16.85 lb/ft³ (top of curve).

Step 3) Calculate the RC. Compare this to the acceptable relative compaction of 95% to determine that it is not sufficient.

$$RC = \frac{\gamma_{d\,field}}{\gamma_{d\,max}^{*}} \cdot 100 = \frac{15.85\ \frac{lb}{ft^3}}{16.85\ \frac{lb}{ft^3}} \cdot 100 = 94\%\ (\ <95\%\ ;\ not\ acceptable)$$

Notes:

- CERM Chapter 35
- It is possible and acceptable to get over 100% on compaction tests. Much higher than around 105% may indicate the material is changing significantly from what was used for the standard Proctor test.
- Remember that the volume of the mold for modified and standard proctor tests is V_{mold}=1/30 ft³.
- Soil with optimum water content does not require as much compaction to reach the specified compaction.

4) A 20 m long structural steel member is used as a slender column to support a load of 325 kN. There are no intermediate supports. One end is pinned, the other is fixed against rotation but free. The required moment of inertia (m^4) is most nearly:

A) 65.1×10^3
B) <u>5.27×10^{-4}</u>
C) 13.0×10^{-5}
D) 13.2×10^{-5}

$E = 20 \times 10^{10}$ Pa
FS = 2

FIND: Moment of inertia, I (m^4)

Step 1) Search for *slender column* in the Handbook and navigate to *Chapter 1.6.8 Columns.* Euler's Formula can be used to solve for the required moment of inertia:

$$P_{cr} = \frac{\pi^2 EI}{(K\ell)^2} \Rightarrow I = \frac{P_{cr}(K\ell)^2}{\pi^2 E}$$

Step 2) Calculate the design load, or maximum load, given the Factor of Safety of 2:

$$P_{cr} = (325\ kN)(2) = 650\ kN = 650\ x\ 10^3\ N$$

Step 3) Find the theoretical effective-length factor, *K*, for end conditions: K=2.0

Step 4) Calculate the required moment of inertia:

$$I = \frac{P_{cr}(K\ell)^2}{\pi^2 E} = \frac{(650\ x\ 10^3\ N)(2 \cdot 20\ m)^2}{\pi^2 \left(20\ x\ 10^{10}\ \frac{N}{m^2}\right)} = 5.27 x 10^{-4}\ m^4$$

Notes:

- CERM Chapter 45

5) Using the Terzaghi-Meyerhof bearing capacity equation and the Terzaghi bearing capacity factors, what is most nearly the ultimate bearing capacity (kPa) of the continuous strip footing shown?

A) 951
B) <u>1,224</u>
C) 1,646
D) 25,701

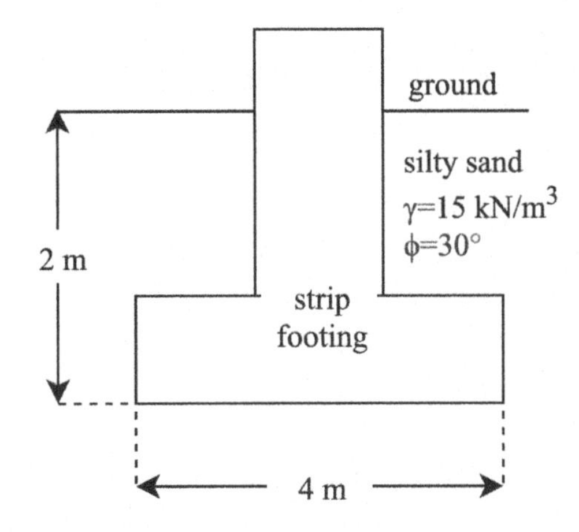

FIND: Ultimate bearing capacity, q_{ult} (kPa)

Step 1) Search for *bearing capacity* in the Handbook and navigate to *Chapter 3.4.2 Bearing Capacity Equation for Concentrically Loaded Strip Footings*:

$$q_{ult}=c(N_c)+(q_{appl}+\gamma_a D_f)(N_q)+0.5(\gamma)(B_f)(N_\gamma)$$

Step 2) Determine the Terzaghi bearing capacity factors for ϕ=30° from the Bearing Capacity Factors table:

$$N_\gamma = 22.4 \qquad N_c = 30.1 \qquad N_q = 18.4$$

Step 2) Using $c=0$ because sand is cohesionless, and $q_{appl}=0$ because there is no net surcharge, the ultimate bearing capacity can be calculated:

$$q_{ult}=c(N_c)+(q_{appl}+\gamma_a D_f)(N_q)+0.5(\gamma)(B_f)(N_\gamma)$$

$$q_{ult}=\left(0\frac{kN}{m^2}\right)(30.1)+\left(0\frac{lbf}{ft^2}+\left(15\frac{kN}{m^3}\right)(2\ m)\right)(18.4)+0.5\left(15\frac{kN}{m^3}\right)(4\ m)(22.4)=1,224\ kPa$$

Notes:

- CERM Chapter 36
- A surcharge pressure would likely be shown as an additional force on the diagram given in the problem or given in the written question.
- Do not expect to be given which equation to use.
- The cohesion of ideal sand is zero. This is a useful thing to remember!

6) A horizontal curve for a highway has the elements shown below. The intersection angle (degrees) is most nearly:

A) <u>55.55</u>
B) 27.78
C) 55.25
D) 56.00

M = 138.10
D = 4.78°
T = 631.35 ft

FIND: Intersection angle, I (degrees)

Step 1) Search for *horizontal curves* in the Handbook and navigate to *Chapter 5.2.1 Basic Curve Elements.*

Step 2) Calculate R:

$$R = \frac{5,729.6 \frac{ft}{deg}}{D} = \frac{5,729.6 \frac{ft}{deg}}{4.78\ deg} = 1,198.66\ ft$$

Step 3) Calculate I:

$$T = R \tan \frac{\Delta}{2} \Rightarrow \Delta = 2 \tan^{-1}\left(\frac{T}{R}\right)$$

$$\Delta = 2 \tan^{-1}\left(\frac{631.35\ ft}{1,198.66\ ft}\right)$$

$$\Delta = 55.55°$$

Notes:

- CERM Chapter 79
- It is generally safe to assume *arc basis* if *chord basis* is not specified. The *arc basis* is most often used unless you are dealing with a very large radii, such as a railroads. This is because railroads have such large radii that some short portions of the curve are essentially straight.
- The given M is simply to confuse the reader. You may encounter excess information and you will have to know what you need and don't need!

7) The effective stress (pcf) at the middle of Layer 3 is most nearly:

A) 2,058.80
B) 2,164.40
C) 2,682.80
D) 15,630.00

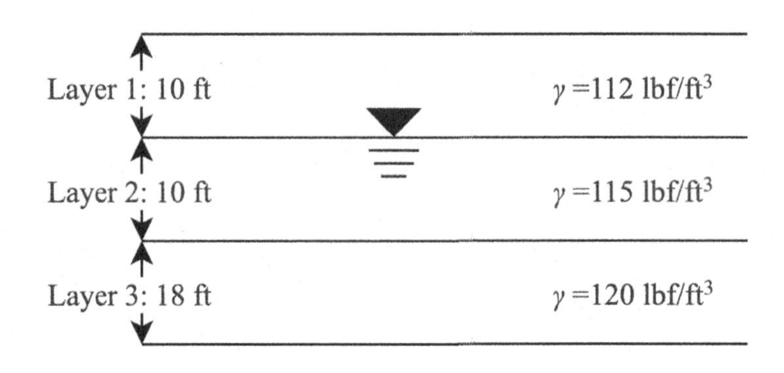

Find: Effective stress, σ' (pcf)

Two ways to solve this are shown here, one far more simple than the other. Sometimes the simpler method is only reliable if you have a solid understanding of the concepts, otherwise it is easy to make mistakes.

Option 1)

Step 1) Calculate the total stress by summing the stress induced by each layer up to the required point:

$$\sigma = \sum \gamma z$$

$$= \left(112 \frac{lbf}{ft^3}\right)(10 ft) + \left(115 \frac{lbf}{ft^3}\right)(10 ft) + \left(120 \frac{lbf}{ft^3}\right)\left(\frac{18}{2} ft\right)$$

$$= 3,350 \frac{lbf}{ft^3}$$

Step 3) Calculate the pore water pressure at the same point:

$$u_z = \gamma_{water} z_{water}$$

$$= \left(62.4 \frac{lbf}{ft^3}\right)\left(10 ft + \frac{18}{2} ft\right) = 1,185.60 \frac{lbf}{ft^2}$$

Step 4) Calculate the effective stress:

$$\sigma' = \sigma - u = 3,350 \frac{lbf}{ft^2} - 1,185.60 \frac{lbf}{ft^2} = 2,164.40 \frac{lbf}{ft^2}$$

Option 2)

Step 1) Directly calculate the effective stress :

$$\sigma' = \sum \sigma - u$$

$$= 10\,ft \cdot 112\,pcf + \left[10ft(115\,pcf - 62.4\,pcf) + \left(\frac{1}{2}\right) 18ft(120\,pcf - 62.4\,pcf) \right]$$

$$= 2,164.40\,pcf$$

Notes:

- CERM Chapter 35
- Remember to solve for the point specified in the problem!
- Although the units are the same, pressure is exerted on a body/fluid, whereas stress is exerted from inside the body.
- The Handbook does not cover effective stress thoroughly, but it is important for you to know that effective stress is the total stress minus the pore water pressure.

8) Which of the following most adequately describes the construction procedure for concrete using the following steps?

A) VI, V, I, VII, IV, III, II
B) VII, I, IV, V, II, VI, III
C) VII, II, III, VI, I, IV, V
D) VII, I, II, III, IV, V, VI

I) Prepare the formwork
II) Pour the concrete
III) Disassemble the formwork
IV) Place steel rebar
V) Prepare concrete mix
VI) Cure the concrete
VII) Preparing the site

Notes:

- General knowledge of the concrete pouring procedure may be useful.

9) The cantilever frame shown is supported by a 10m x 10m footing. Neglect the thickness and weight of the footing. The maximum and minimum tensile stress equations are given below. The maximum normal tensile stress (kN/m^2) below the footing is most nearly:

A) 0.204
B) <u>0.404</u>
C) 1.084
D) 1.087

$$\sigma_{max,\ min} = \frac{F}{A} \pm \frac{Mc}{I_c}$$

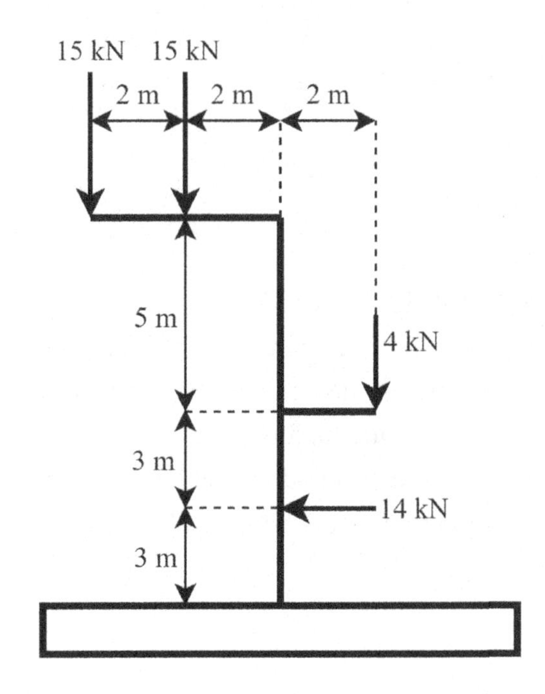

Find: Maximum normal tensile stress (kN/m^2)

Step 1) The bending moment at the foundation level must be calculated first because normal stress occurs as a combined effect of both axial forces and bending moments. Remember to use the convention that the force, F, is negative if the force compresses the member.

$$F = F_{axial}$$

$$= 15\ kN + 15\ kN + 4\ kN = 34\ kN = -34\ kN\ (\ compression)$$

$$M = \sum moments = \sum F \cdot d$$

$$= (\ 15\ kN)\ (\ 2\ m) + (\ 15\ kN)\ (\ 4\ m) + (\ 14\ kN)\ (\ 3\ m) - (\ 4\ kN)\ (\ 2\ m)$$

$$= 124\ kN \cdot m$$

Step 2) Calculate the maximum normal tensile stress:

$$\sigma = \frac{F}{A} \pm \frac{M}{S} = \frac{F}{A} \pm \frac{M}{\left(\frac{bh^2}{6}\right)}$$

$$= \frac{-34\ kN}{(10\ m)(10\ m)} \pm \frac{124\ kN \cdot m}{\left(\frac{(10\ m)(10\ m)^2}{6}\right)}$$

$$= +0.404 \frac{kN}{m^2} (\text{tension})$$

$$-1.084 \frac{kN}{m^2} (\text{compression})$$

Notes:

- CERM Chapter 44

10) Water flows at 135 L/s through a schedule-40 steel pipe that gradually enlarges in diameter from point A to point B. All minor losses are insignificant. The total energy at point B (J/kg) is most nearly:

A) 264
B) 265
C) 291
D) 301

Point A
Elevation = 456 meters
$P_A = 70$ kPa
$E_{A, total} = 94.50$ J/kg

Point B
Elevation = 478 meters
$P_B = 49$ kPa
$D_B = 0.50$ meters

FIND: Total energy at point B, $E_{B, total}$ (J/kg)

Step 1) Search *energy equation* in the Handbook and navigate to *Chapter 6.2.1.3 Field Equation.* Each side of the equation shown should be recognizable to you as the total energy at that point (interchanging "2" in the equation for "B" in the problem):

$$E_{t,B} = E_p + E_v + E_z = \frac{p_B}{\rho} + \frac{v_B^2}{2} + z_B g$$

Step 2) Calculate the velocity at point B:

$$Q = v_B A_B = v_B \left(\frac{\pi}{4}\right) D_B^2 \qquad \Rightarrow v_B = \left(\frac{4}{\pi}\right)\left(\frac{Q}{D_B^2}\right) = \left(\frac{4}{\pi}\right) \frac{\left(135 \frac{L}{s}\right)}{(0.50\ m)^2 \left(1{,}000 \frac{L}{m^3}\right)} = 0.69 \frac{m}{s}$$

Step 3) Calculate the total energy at point B:

$$E_{t,B} = E_p + E_v + E_z = \frac{p_B}{\rho} + \frac{v_B^2}{2} + z_B g$$

$$= \frac{49{,}000\ Pa}{1{,}000 \frac{kg}{m^3}} + \frac{\left(0.69 \frac{m}{s}\right)^2}{2} + (22\ m)\left(9.81 \frac{m}{s^2}\right) = 265.06 \frac{J}{kg}$$

Notes:

- CERM Chapter 16
- You will also see Q denoted as $\dot{V}$. Both are used for the volumetric flow rate depending on the application.

11) A slope is shown below with soil parameters. Most nearly, what is the Factor of Safety against shear failure using the Taylor method?

A) 0.95
B) <u>1.74</u>
C) 1.76
D) 7.07

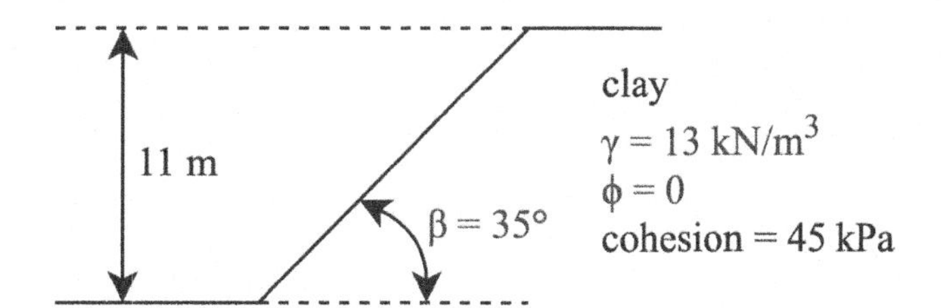

FIND: Factor of Safety, F

Step 1) Search *factor of safety* in the Handbook and navigate to *Chapter 3.6.1 Stability Charts*. Within the Stability Number chart is the equation for the Factor of Safety:

$$Factor\ of\ Safety\ F = N_o \frac{c}{\gamma H}$$

Step 2) Because the depth of the firm stratum is not given, assume an infinite depth from the surface:

$$d = \frac{D}{H} = \frac{\infty}{11\ m} = \infty$$

Step 3) Because d=∞, deduce from the Stability Number chart that the dimensionless stability number, $N_o = 5.53$.

Step 4) Calculate the Factor of Safety:

$$Factor\ of\ Safety\ S = N_o \frac{c}{\gamma H} = (5.53)\frac{(45\ kPa)}{\left(13\frac{kN}{m^3}\right)(11\ m)} = 1.74$$

Notes:

- CERM Chapter 40
- Know what to do if the soil is saturated or submerged. If no water table is present: $\gamma_{total} = \gamma_{effective}$
- Generally, the minimum factor of safety is around 1.3-1.5, so be aware if you get something lower than this.

12) The following information is submitted for a proposed concrete mix. The absolute volume of fine aggregate (ft^3) in 1 yd^3 of fresh concrete is most nearly:

A) 3.40
B) 4.11
C) <u>9.52</u>
D) 17.48

Cement	
Specific gravity	3.10
Cement content	7 sack mix
Fine aggregate	
Specific gravity	2.48
Fineness modulus	2.75
Absorption	3.5%
Coarse aggregate	
Volume	8.96 ft^3/yd^3
Specific gravity	2.70
Dry	1,418 lbf/yd^3
Water-Cement Ratio	0.39
Air content	3.75%

FIND: Absolute volume of fine aggregate, $V_{aggregate}$ (ft^3)

Step 1) Determine the volume of all other components:

Ingredient	Weight (lbf per cubic yard)	Volume (ft^3 per cubic yard)
Cement	$(7\ sacks)\left(94\dfrac{lbf}{sack}\right)=658\ lbf$	$\dfrac{658\ lbf}{(3.10)\left(62.4\dfrac{lbf}{ft^3}\right)}=3.40\,ft^3$
Coarse aggregate	-	8.96 ft^3 (given in problem)
Air	0	$(0.0375)\left(1\ yd^3\right)\left(27\dfrac{ft^3}{yd^3}\right)=1.01\,ft^3$
Water	(0.39)(658 lbf) = 256.62 lbf	$\dfrac{256.62\ lbf}{62.4\dfrac{lbf}{ft^3}}=4.11\,ft^3$
	Total:	17.48 ft^3

Step 2) Calculate the volume of fine aggregate in 1 yd^3 of the mix:

$$\left(1\ yd^3\right)\left(27\frac{ft^3}{yd^3}\right)-17.48\,ft^3=9.52\,ft^3$$

Notes:

- CERM Chapter 49
- Remember that on concrete mix problems, all you are working with is cement, air, water, and aggregate. Therefore, it is important you understand how to find the weights and volumes of these from what is given.
- Concrete is generally measured, sold, made, and poured in cubic yards. This means that generally all of your calculations are *per cubic yard*, even if that is not written.
- Also know how to adjust for *saturated surface dry* (SSD) conditions, which occurs when the aggregate holds as much water as it can without trapping any free water between the aggregate particles. The necessary equations for these adjustments can be found in *Chapter 5.5.5.3 Specific Gravity of Fine Aggregates*. No adjustments need to be made for the concrete (only the aggregate).
- Remember:
    - 1 gallon of water = 8.34 lbs
    - 1 ft^3 of water = 62.4 lbs
    - 1 sack of cement = 94 lbs

13) Sewer and water pipes are designed to have 18 in. of vertical separation (outside-of-pipe to outside-of-pipe). The sewer main has been installed per design with a flowline elevation of 4782.28 ft at a location where a water main crosses above the sewer. The water and sewer pipes both have an inner diameter of 12 in. and outer diameter of 13.75 in. The rod reads 10.76 ft on the top of the water main. Assuming a flat existing ground, the ground elevation (ft) directly at the top of the trench is most nearly:

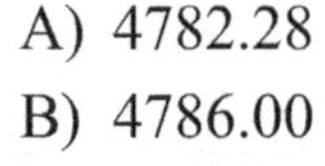

A) 4782.28
B) 4786.00
C) <u>4793.26</u>
D) 4796.76

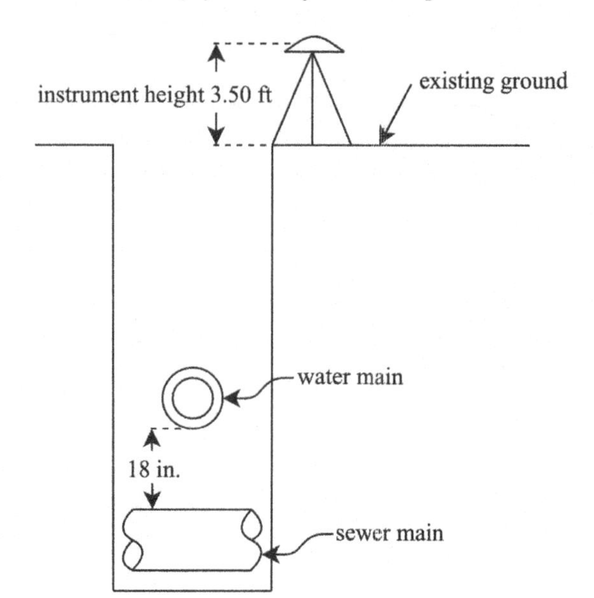

FIND: Ground elevation (ft)

Step 1) With an inner and outer diameter of 12 in. and 13.75 in., respectively, the pipe thickness is 0.875 in. Calculate the ground elevation:

$$ground\ elevation = 4782.28\,ft + \frac{12\,in. + 0.875\,in.}{\frac{12\,in.}{ft}} + \frac{18\,in.}{\frac{12\,in.}{ft}} + \frac{13.75\,in.}{\frac{12\,in.}{ft}} + 10.76\,ft - 3.50\,ft$$

$$= 4793.26\,ft$$

14) Settling is generally due to consolidation of the supporting soil. Which are the three distinct periods of consolidation?

A) Primary, secondary, and tertiary
B) <u>Immediate, primary, and secondary</u>
C) Immediate, overconsolidation, and normal
D) Vertical, horizontal, and secondary

Notes:

- CERM Chapter 40
- Although the Handbook does not cover all three distinct periods of consolidation, the three distinct periods of consolidation are immediate, primary, and secondary.

15) An 8 ft long cantilever beam is loaded by 250 lbf at 2 ft from its free end. The cross section is 8 in. wide by 4 in. high. The maximum deflection of the beam is most nearly:

A) 0.24
B) 0.49
C) <u>0.73</u>
D) 42.7

$E = 1.5 \text{ x } 10^6$ psi

FIND: Total dip deflection, $y_{total\ at\ tip}$ (inches)

Step 1) Look up *deflection* in the Handbook and navigate to the table in Chapter 1 entitled Cantilevered Beam Slopes and Deflections. Find the equation for max deflection:

$$v_{max} = \frac{-Pa^2}{6EI}(3L - a)$$

Step 2) Find the moment of inertia of the beam cross section in the table in *Chapter 1.5.12 Concurrent Forces.*

$$I = \frac{bh^3}{12} = \frac{(8\ in)(4\ in)^3}{12} = 42.67\ in^4$$

Step 3) Calculate the deflection at the 250 lbf load:

$$v_{max} = \frac{-Pa^2}{6EI}(3L - a) = \frac{-(250\ lbf)(96\ in\ -\ 24\ in)^2}{6\left(1.5\ x\ 10^6 \frac{lbf}{in^2}\right)(42.67\ in^4)}(3 \cdot 96\ in\ -\ 72\ in) = -0.73\ in\ (0.73\ in)$$

Notes:

- CERM Chapter 44

16) A bridge is proposed to cross over the sag vertical curve shown below. If 22 ft of clearance is required, what is most nearly the lowest elevation (ft) that the bottom of the overhead tram bridge may have at station 22+20?

A) 698.35
B) 698.41
C) 712.43
D) <u>724.53</u>

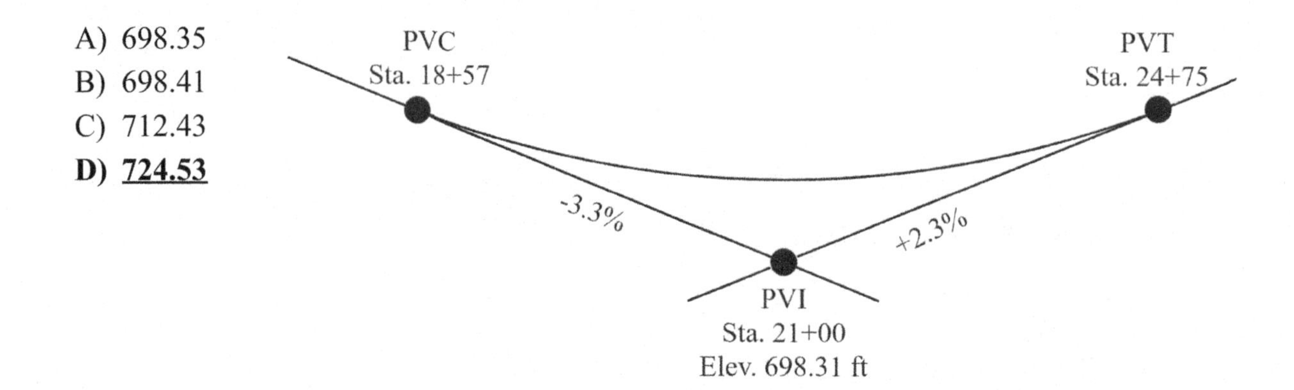

FIND: Lowest allowable elevation at Station 22+20 (ft)

Step 1) Search *vertical curve* in the Handbook and navigate to *Chapter 5.3.1 Symmetrical Vertical Curve Formula.*

Step 2) Determine the length of the curve, L:

$$L = (Sta.24+75) - (Sta.18+57) = 6.18\ stations$$

Step 3) Solve for the rate of change per station, *r*:

$$r = \frac{(g_2 - g_1)}{L} = \frac{(2.3\% - (-3.3\%))}{6.18\ stations} = 0.91 \frac{\%}{station}$$

Step 4) Calculate the elevation at PVC:

$$elev_{PVC} = elev_{PVI} - G_1\left(\frac{L}{2}\right)$$

$$= 698.31\,ft - (-3.3\%)\left(\frac{6.18\ stations}{2}\right) = 708.51\,ft$$

Step 5) Calculate the distance from PVC to Sta. 22+20, $x_{Sta.\ 22+20}$:

$$x_{22+20} = (Sta.22+20) - (Sta.18+57) = 3.63\ stations$$

Step 6) Use the given curve elevation equation to calculate the elevation at Sta. 22+20:

$$curve\ elevation = Y_{PVC} + g_1 x + ax^2 = Y_{PVC} + g_1 x + x^2\left(\frac{g_2 - g_1}{2L}\right)$$

$$elev_{22+20} = 708.51\,ft + (-3.3\%)\,(3.63\ stations) + \frac{\left(0.91\frac{\%}{station}\right)(3.63\ stations)^2}{2}$$

$$= 702.53\,ft$$

Step 7) Calculate the allowable elevation of the bottom of the tram bridge:

$$elev_{bottom\ of\ tram\ bridge} = elev_{22+20} + 22\,ft = 702.53\,ft + 22\,ft = 724.53\,ft$$

Notes:

- CERM Chapter 79
- Note that g is a decimal and L is the length of curve in feet. You can either use those units, or you can keep g in percent and make L in stations as is shown in this problem. Be careful not to mix and match though!
- Do not mix up a specific station (e.g. Station 22+20) with general station amounts (e.g. x_{20+20} is 3.63 stations past the PVC). We tried to make this clear here by specifying Station 22+20 versus 3.63 stations or 0.91%/station. You may just see "sta" applied to both in other references.

17) An 8.5 ft tall retaining wall to be made of normal weight concrete is being poured on a day with a measured temperature of 92° F. The concrete mixture is Type I with no added retarders. The concrete lateral pressure distribution equation is given below. The reduction in hydrostatic pressure (psf) due to pour and set up rates is most nearly:

A) 541
B) 600
C) <u>675</u>
D) 3,000

$$p_{max,psf} = 600C_w \le C_w C_c\left(150 + \frac{9000R_{ft/hr}}{T_{°F}}\right) \le 150h_{ft}$$

$C_w = 1.0$
$C_c = 1.0$
$R = 4$ ft/hr
$\gamma = 150$ lbf/ft^3

FIND: Pressure reduction, ΔP (psf)

Step 1) Determine the minimum and maximum pressures using the given equation:

$$p_{min} = C_w\left(600\,\frac{lbf}{ft^2}\right) = (1.0)\left(600\,\frac{lbf}{ft^2}\right) = 600\frac{lbf}{ft^2}$$

$$p_{max} = 150\,h_{ft} = (150)\,(8.5\,ft) = 1,275\,\frac{lbf}{ft^2}\left(< 2,000\frac{lbf}{ft^2}\;OK\right)$$

$$p_{calc} = C_w C_c\left(150\,\frac{lbf}{ft^3} + 9,000\left(\frac{R_{ft/hr}}{T_{°F}}\right)\right) = (1.0)\,(1.0)\left(150\,\frac{lbf}{ft^3} + 9,000\left(\frac{4\,\frac{ft}{hr}}{92\,°F}\right)\right) = 541\frac{lbf}{ft^2}$$

Step 4) Ensure you use the correct values:

$$p_{calc} \le p_{min}$$

Therefore, use:

$$p_{design} = p_{min} = 600\,\frac{lbf}{ft^2}$$

Step 5) Calculate the difference between the hydrostatic pressure without pouring and setup rates, and the hydrostatic pressure with the pouring and setup rates:

$$\Delta p = p - p_{max,psf} = 1,275\frac{lbf}{ft^2} - 600\frac{lbf}{ft^2} = 675\frac{lbf}{ft^2}$$

Notes:

- CERM Chapter 49
- If C_w and C_c are not given and cannot be figured out from information given, it is generally safe to assume a value of 1.0.
- These equations are dimensionally inconsistent.
- If you're studying with the CERM, this is the only topic that I found in my studies that differs between the 14th and 15th CERM Editions (this does not mean it's the only one). This topic is found on CERM 49-8.16 *Lateral Pressure on Formwork* (14th Ed.), but you will see that the equations do not contain the C_w or C_c (unit weight and chemistry coefficients, respectively).

18) A construction site near a waterway requires temporary best management practices (BMPs) for erosion and sediment. A slope drain is used to control the flow of water off of the construction site. Which of the following is most appropriate to control erosion at the end of the slope drain?

A) Sediment structure
B) Temporary berm
C) Silt fence
D) Temporary seeding

You should have a basic understanding of best management practices for erosion and sediment. A *sediment structure* is an energy-dissipating structure. Silt fences should not receive concentrated flow and therefore would not be appropriate. A temporary berm could become eroded and contribute to the problem. Temporary seeding does not dissipate energy.

19) \$550 is compounded monthly at a 5% nominal annual interest rate. Most nearly how much (\$) will have accumulated in 4 years?

A) 450
B) 669
C) <u>672</u>
D) 2,875 **FIND: Future amount, F (\$)**

Step 1) Search *economics* in the Handbook and navigate to *Chapter 1.7 Engineering Economics.* Draw a diagram representing F given P (F/P):

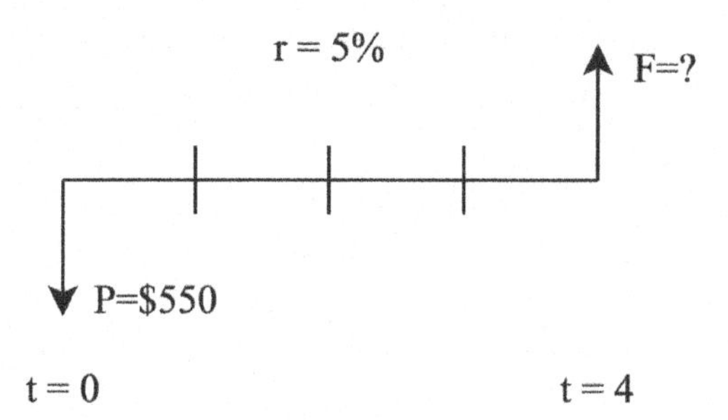

Step 2) Calculate the interest rate from *Chapter 1.7.2 Nonannual Compounding*:

$$i = \left(1 + \frac{r}{m}\right)^m - 1 = \left(1 + \frac{0.05}{12}\right)^{12} - 1$$
$$= 0.0512$$
$$= 5.12\%$$

Step 3) Calculate the future amount, F, using the F/P formula from the factor table:

$$F = P(1+i)^n = (\$550)(1+0.0512)^4 = \$671.59$$

Notes:
- CERM Chapter 87

20) Primary consolidation occurs in clayey soils due primarily to:

A) <u>Extrusion of water from the void spaces</u>
B) Plastic readjustment of the soil grains
C) Compression of the soil grains
D) Specific gravity of the soil

Notes: CERM Chapter 40

21) The following volumes shown are the cross sectional areas at the respective stations along a roadway. Most nearly, what is the total volume excavated (yd^3)?

A) 4,900
B) <u>4,919</u>
C) 132,795
D) 14,755

Station	Area (sf)
12+50	0
12+96	510
13+25	234
15+46	528
15+55	498
16+01	435

FIND: Volume, V (yd^3)

Step 1) Search *earthwork* in the Handbook and navigate to the average end-area method in *Chapter 2.1.2.1*:

$$V = L\frac{(A_1 + A_2)}{2}$$

Step 2) Calculate the total volume excavated for each stretch:

$$V_{12+50 \rightarrow 12+96} = 46\,ft\,\frac{(0\,ft^2 + 510\,ft^2)}{2} = 11,730\,ft^3$$

$$V_{12+96 \rightarrow 13+25} = 29\,ft\,\frac{(510\,ft^2 + 234\,ft^2)}{2} = 10,788\,ft^3$$

$$V_{13+25 \rightarrow 15+46} = 221\,ft\,\frac{(234\,ft^2 + 528\,ft^2)}{2} = 84,201\,ft^3$$

$$V_{15+46 \rightarrow 15+55} = 9\,ft\,\frac{(528\,ft^2 + 498\,ft^2)}{2} = 4,617\,ft^3$$

$$V_{15+55 \rightarrow 16+01} = 46\,ft\,\frac{(498\,ft^2 + 435\,ft^2)}{2} = 21,459\,ft^3$$

Step 3) Sum the total excavated volume:

$$\sum_V = 11,730\,ft^3 + 10,788\,ft^3 + 84,201\,ft^3 + 4,617\,ft^3 + 21,459\,ft^3 = 132,795\,ft^3$$

$$132,795\,ft^3\left(\frac{1\,yd}{3\,ft}\right)^3 = 4,918.33\,yd^3$$

Notes: CERM Chapter 80

22) A frictionless retaining wall is shown. The resultant of the active earth pressure (kN) acting on the wall is most nearly:

A) 37
B) 99
C) 101
D) 168

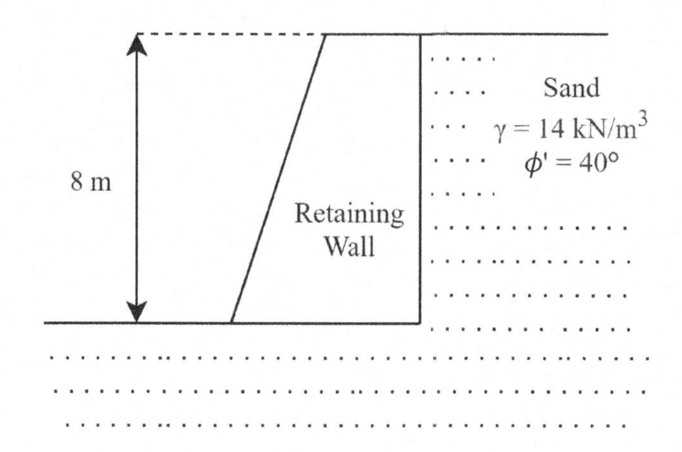

FIND: Resultant of active earth pressure, R_a (kN)

Step 1) Search *retaining wall* in the Handbook and navigate to the diagram entitled Failure Surfaces, Pressure Distribution and Forces: (a) Active case, (b) Passive case in *Chapter 3.1.2 Rankine Earth Coefficients*. Use the equation for Active force within depth z:

$$P_a = \frac{K_a \gamma z^2}{2}$$

Step 2) Visualize that the sand is exerting a force on the retaining wall in the shape of a triangle:

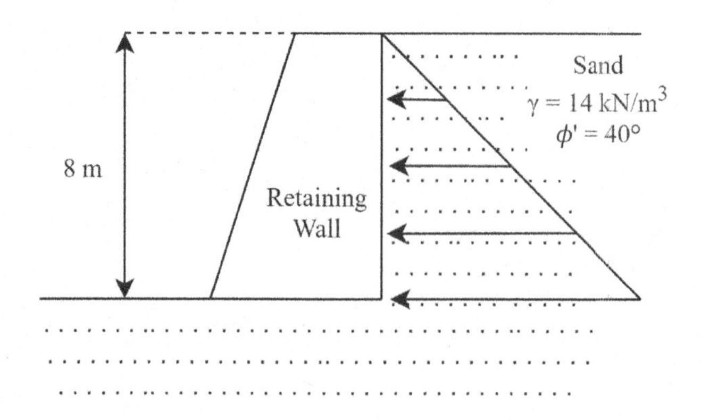

Step 3) Calculate the coefficient of active earth pressure:

$$K_a = \frac{1 - \sin\phi'}{1 + \sin\phi'} = \frac{1 - \sin(40°)}{1 + \sin(40°)} = 0.22$$

Step 4) Calculate the resultant of the active earth pressure:

$$P_a = \frac{K_a \gamma z^2}{2} = \frac{(0.22)\left(14\frac{kN}{m^3}\right)(8\ m)^2}{2} = 98.56\ kN$$

Notes:

- CERM Chapter 37
- It's common to round on K values, but notice how different the answer is if you use the unrounded K value. Although we round here, it is always safe to store the unrounded value and use throughout the calculations. The solutions on the exam should be far enough apart to account for rounding differences!

23) Which following statement regarding transportation Levels of Service is most accurate?

A) Level A represents severely impeded flow.
B) Level B generally has the maximum capacity.
C) Economic considerations favor less obstructed levels of service.
D) Political considerations favor less obstructed levels of service.

Notes: CERM Chapter 73

24) The short column shown carries a maximum concentric load of 200 kN and has a square cross section. The allowable strength of the column material is 200 kN/m^2. The allowable stress can be equal to or greater than the actual stress. The minimum required side length (m) of the square is most nearly:

A) <u>1.00</u>
B) 1.10
C) 2.00
D) 2.10

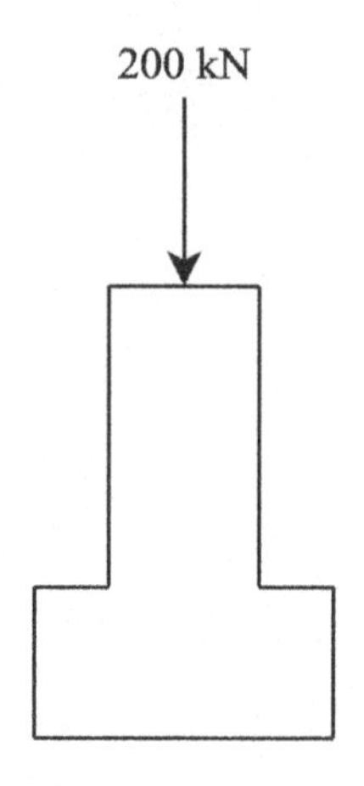

FIND: Minimum side length, b (meters)

Step 1) Stress is equal to force divided by area:

$$\sigma_{actual} = \frac{P}{A} = \frac{200\ kN}{b^2}$$

$$\sigma_a = 200\ \frac{kN}{m^2}$$

$$\frac{200\ kN}{b^2} \le 200\ \frac{kN}{m^2}$$

$$b = 1\ m$$

Notes:

- CERM Chapter 45
- Be very familiar with stress, its units, and how to calculate it.

25) Which statement is false?

A) The Bernoulli equation states that the sum of the pressure, velocity, and elevation heads is constant.
B) The *hydraulic grade line* depicts potential energy (position plus pressure head) at all stations along the system.
C) The *velocity head* is equal to the elevation of the *energy grade line* minus the elevation of the *hydraulic grade line.*
D) In the Bernoulli Equation, the difference between the hydraulic grade line and the energy line is the *P/γ* term.

Search *Bernoulli* and navigate to *Chapter 6.2.1.5 Energy Line* (Bernoulli Equation).

Notes:
- CERM Chapter 16
- Taking a moment to soak in the correct answers can help your understanding during your studies, and could surprise you by helping on other problems during the exam!

26) The mechanical properties of structural steel are generally determined by what kind of tests?

A) Tension
B) Compression
C) Torsion
D) Heat

Notes: CERM Chapter 58

27) Which type of reinforcement is appropriate for concrete that is exposed to weather or in contact with the ground?

A) No. 14 and No. 18 bars
B) No. 6 through No. 18 bars
C) No. 11 bar and smaller
D) Stirrups, ties, and spirals

Search *reinforcement* in the Handbook and navigate to *Chapter 2.5.2.3 Concrete Cover Requirements for Reinforcement.* Find the correct answer in the first table in this section.

28) Soil extracted from a site at a depth of 15 m has an Atterberg liquid limit of 45.2 and plastic limit of 14.9. The grain size distribution is shown below. What is the classification of the soil according to the AASHTO system?

A) A-7-6
B) A-6
C) A-2-5
D) A-8

US sieve size	Cumulative percentage retained
No. 4	0%
No. 10	7.1%
No. 40	18%
No. 200	36.7%

FIND: Soil Classification

Step 1) Search *atterberg* in the Handbook and navigate to *Chapter 3.8.2 Atterberg Limits.*

Step 2) For each sieve, subtract the cumulative percentage retained from 100% to get the percentage passing.

US seive size	Cumulative percentage passing
No. 4	100% - 0% = 100%
No. 10	100% - 7.1% = 92.9%
No. 40	100% - 18% = 82%
No. 200	100% - 36.7% = 63.3%

Step 3) Calculate the plasticity index:

$$PI = LL - PL = 45.2 - 14.9 = 30.3$$

Step 4) Now search *AASHTO Classification* in the Handbook and navigate to the table in AASHTO Classification System. Automatically sort this soil into the silt-clay materials because the soil has more than 35% passing the no. 200 sieve. Moving down the table, we can determine that the soil is A-7 because the LL (45.2) is above the 41 minimum, and the PI (30.3) is above the 11 minimum.

Step 5) Look at the table subcaptions to determine that the soil is classified as A-7-6 because the PL (14.9) is less than 30.

Notes:

- CERM Chapter 35
- Always pay attention to *retained* versus *passing* because the AASTHO Soil Classification table uses *passing*.
- To move more quickly through this table, you can immediately split the soil sample into *granular* and *silt-clay* materials based on the top row showing percent passing no. 200 sieve.
- Pay attention to *min* and *max* on the table. It is easy to look too quickly and look at the wrong one.

29) A steel layout plan is shown including the section lengths and weights. The total weight (lbs) of the steel is most nearly:

A) 33,145
B) 38,160
C) 51,475
D) <u>52,225</u>

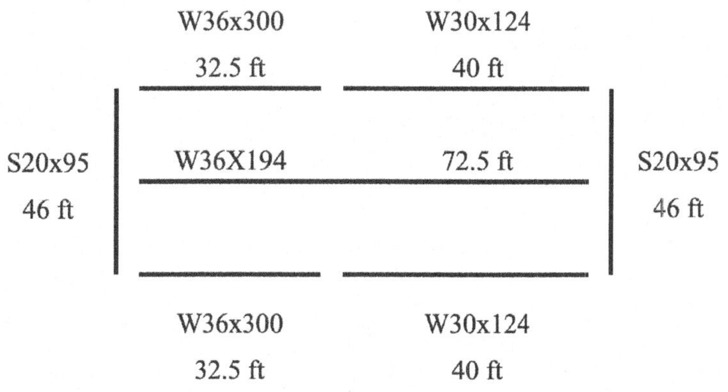

FIND: Steel weight (lbs)

Step 1) The second number of the steel specifications is the weight in lbs/ft. Find the weight of the steel:

$$\text{S20x95: } 95\ lbs/ft\ x\ (46ft\ x\ 2) = 8,740\ lbs$$
$$\text{W36x300: } 300\ lbs/ft\ x\ (32.5ft\ x\ 2) = 19,500\ lbs$$
$$\text{W30x124: } 124\ lbs/ft\ x\ (40ft\ x\ 2) = 9,920\ lbs$$
$$\text{W36x194: } 194\ lbs/ft\ x\ 72.5\ ft = 14,065\ lbs$$
$$8,740\ lbs + 19,500\ lbs + 9,920\ lbs + 14,065\ lbs = 52,225\ lbs$$

30) An 85-ac watershed has two distinct soil types shown below. The storage capacity of the soil (in.) is most nearly:

A) 0.90
B) 1.30
C) <u>3.80</u>
D) 11.30

24 acres Straight row broadcast legumes Poor hydrologic condition I = 0.35 in/hr	61 acres Straight row small grain Good hydrologic condition I = 0.16 in/hr

FIND: Storage capacity, S (inches)

Step 1) Search *NRCS* in the Handbook and navigate to *Chapter 6.5.2.2 NRCS (SCS) Rainfall Runoff Method.* Find the equation for S, maximum basin retention (in.):

$$S = \frac{1000}{CN} - 10$$

Step 2) Determine the hydrologic soil group according to its infiltration rate from the table entitled Minimum Infiltration Rates for the Various Soil Groups:

The 24-ac area with an infiltration rate of 0.35 in/hr falls in Group A.

The 61-ac area with an infiltration rate of 0.16 in/hr falls in Group B.

Step 3) Determine the curve numbers, CN, from the Runoff Curve Numbers table:

$$CN_{24\text{-acre}} = 66$$
$$CN_{61\text{-acre}} = 75$$

Step 4) Calculate the composite CN:

$$CN = \frac{(24\ ac)(66) + (61\ ac)(75)}{85\ ac} = 72.46$$

Step 5) Calculate the soil storage capacity:

$$S_{in} = \frac{1000}{CN} - 10 = \frac{1000}{72.46} - 10 = 3.80\ in$$

Notes:

- CERM Chapter 20
- The table in *Chapter 6.5.2.2 NRCS (SCS) Rainfall Runoff Method* is incorrect. The correct infiltration rates for the hydrologic soil groups are: Group A: >0.30 in/hr, Group B: 0.15-0.30 in/hr, Group C: 0.05-0.15 in/hr, and Group D: 0.05 in/hr.

31) Which is the best rebar reinforcement for the retaining wall shown below?

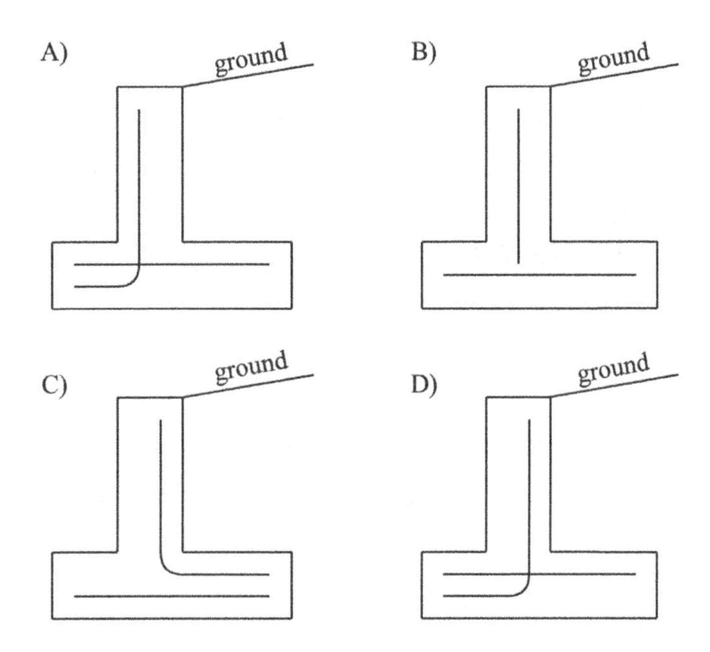

The correct answer is D. Reinforcing bars, or rebar, are placed in concrete because of its tensile properties and its ability to resist tensile forces. Placing it on the compression side adds little value. Additionally, rebar is overlapped for increased strength.

In the diagrams shown, as the ground pushes outward on the retaining wall from the right, the right side of the wall is in tension and the left side would be in compression.

32) The void ratio of the moist, in situ soil sample with the following characteristics is most nearly:

A) 0.31
B) 0.44
C) <u>0.94</u>
D) 1.96

Mass:	58.72 lb (as sampled)
	49.63 lb (after oven drying)
Volume:	0.60 ft^3 (as sampled)
Specific gravity of solids:	2.59

FIND: Void ratio, e

<u>Option 1</u>

Step 1) Search *void ratio* and navigate to the Volume and Weight Relationship table in *Chapter 3.8.3 Weight-Volume Relationships*. See what parameters you are given and what equation you could use from this table. The problem gives the total volume, V, so the following equation can be used:
$e = (V/V_s) - 1$

Step 2) Solve for V_s (equation given in table):

$$V_s = \frac{W_s}{G\gamma_w} = \frac{49.63\ lb}{(2.59)\left(62.4\frac{lb}{ft^3}\right)} = 0.31\,ft^3$$

Step 3: Solve for the void ratio, *e*:

$$e = \frac{V}{V_s} - 1 = \frac{0.60\,ft^3}{0.31\,ft^3} - 1 = 0.94$$

<u>Option 2</u>

Step 1) Calculate V_s as shown above.

Step 2) Solve for V_w:

$$W_w = W_t - W_s = 58.72\ lb - 49.63\ lb = 9.09\ lb$$

$$V_w = \frac{W_w}{\gamma_w} = \frac{9.09\ lb}{62.4\frac{lb}{ft^3}} = 0.15\,ft^3$$

Step 3) Solve for V_g:

$$V_a = V - \left(V_s + V_w\right) = 0.60\,ft^3 - \left(0.31\,ft^3 + 0.15\,ft^3\right) = 0.14\,ft^3$$

Step 4) Solve for void ratio:

$$e = \frac{V}{V_s} = \frac{V_g + V_w}{V_s} = \frac{0.14\,ft^3 + 0.15\,ft^3}{0.31\,ft^3} = 0.94$$

Notes:

- CERM Chapter 35
- The Volume and Weight Relationship table has a variety of equations depending on what parameters you are given. Be able to mix and match!

33) Which of the following statements is false regarding activity-on-node networks?

A) **A critical activity is one without zero total float.**
B) Delaying the starting time of an activity on the critical path will delay the entire project.
C) Float time is the amount of time that an activity can be delayed without affecting the overall schedule.
D) Float time is defined as float = late start - early start

Search *activity-on-node* and navigate to *Chapter 2.4.1.1 CPM Precedence Relationships*, and read the definitions.

Notes:

- CERM Chapter 86
- For problems that ask what answer is true, it is tempting to find the first truest answer and move on. Avoid this temptation and, if possible, prove all the others wrong before you move on. You may find one that appears true, too, in which case you will have to decipher which one is the most correct.
- Remember that the *critical path* is the longest (most days) continuous path of activities through a project, has the lowest *total float* for the project, and determines the date of project completion.

34) It is required that 10 ft of separation be maintained between the edge of the house foundation and the top of the trench slope. If the underlying soil has the type and depths shown, the distance (ft) between the house foundation and bottom of the 6.5 ft deep trench is most nearly:

A) 12
B) 14
C) 15
D) <u>18</u>

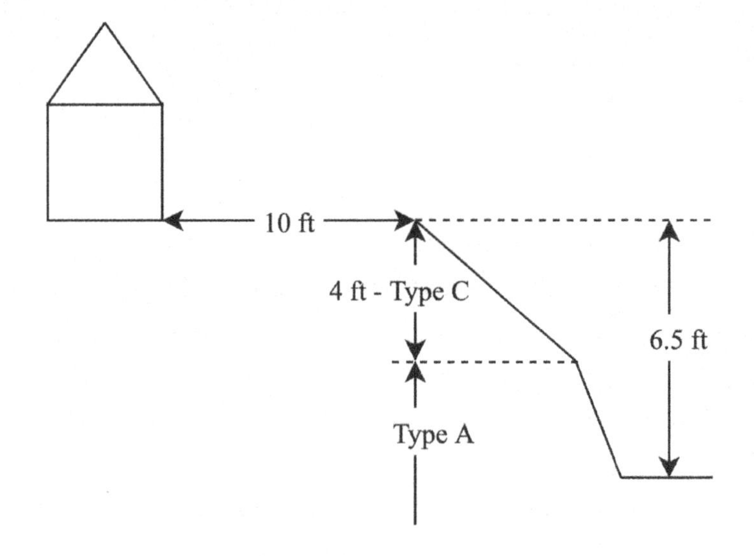

FIND: Distance between house and trench (ft)

Step 1) Search *excavation* in the Handbook and navigate to *Chapter 3.10.3 Slope Configurations: Excavations in Layered Soils.*

Step 2) According to the chart shown, Type C soil over Type A soil requires the Type A soil to be sloped at ¾:1 (run to rise) slope and the Type C soil to be sloped at 1.5:1. Draw distances on the diagram.

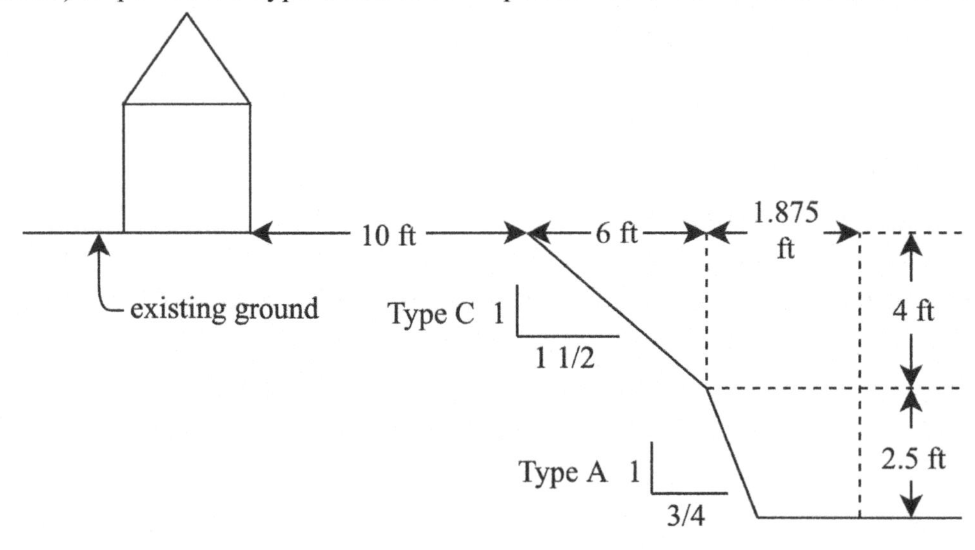

Step 3) Calculate the distance from the house to the bottom of trench:

$$\textit{Distance from house} = 10\,ft + (4\,ft \times 1.50) + (2.50\,ft \times 0.75) = 17.88\,ft$$

Notes: CERM Chapter 83

35) Which of the following is used in determining the experience modification rate?

A) Secondary losses
B) Unpaid losses
C) Reserved losses
D) Unexpected losses

Search experience modification rate and navigate to *Chapter 2.6.1.2 Experience Modification Rate.*

36) The probability of flooding is 1% in any given year. The standard tenure of a facility manager is 30 years. During the 60-year design lifetime of a facility, the probability of flooding is most nearly:

A) 0.40
B) 0.45
C) 0.55
D) 0.56

FIND: Probability of flooding, P

Step 1) Search *flood* and navigate to *Chapter 6.5.1.3 Risk (Probability of Exceedance)*:

$$P\{x \geq x_T \text{ at least once in } n \text{ years}\} = 1 - \left(1 - \frac{1}{T}\right)^n$$

$$P\{100 \text{ year flood in } 60 \text{ years}\} = 1 - \left(1 - \frac{1}{100}\right)^{60 \text{ years}} = 0.45$$

Notes: CERM Chapter 20

37) Most nearly, what is the entrance velocity (ft/s) of a corrugated metal culvert with no headwall and an entrance head loss of 9 ft?

A) 12
B) 20
C) 25
D) 89

FIND: Entrance velocity (ft/s)

Step 1) Search *entrance head loss* in the Handbook and navigate to*Chapter 6.1.10.2 Inlet and Outlet Control.* Find the equation for entrance head loss and rearrange to solve for velocity:

$$H_e = k_e\left(\frac{v^2}{2g}\right) \Rightarrow v = \sqrt{\frac{H_e 2g}{k_e}}$$

Step 2) Search *entrance loss coefficients* in the Handbook and navigate to the table in Chapter 6 entitled Entrance Loss Coefficients. Deduce from the table that a corrugated metal culvert with no headwall has an entrance loss coefficient, $K_e = 0.9$.

Step 3: Calculate the velocity:

$$v = \sqrt{\frac{H_e 2g}{k_e}} = \sqrt{\frac{(9ft)(2)\left(32.2\frac{ft}{s^2}\right)}{0.9}} = 25.38\frac{ft}{s}$$

38) A retention pond is sized to hold 1,600 gallons of water. If a storm produces a runoff flow of 0.56 cfs that is directed into the pond, most nearly how many minutes will the pond take to fill?

E) 0.2
F) 6
G) 8
H) 119

FIND: time, t (minutes)

Step 1) Convert gallons to cubic feet, and solve for time:

$$1,600\ gallons\left(\frac{1\ ft^3}{7.48\ gallons}\right) = 213.90\ ft^3$$

$$t = \frac{V}{Q} = \frac{213.90\ ft^3}{0.56\ \frac{ft^3}{s}\left(\frac{60\ s}{1\ min}\right)} = 6.37\ min$$

Notes:

- Remember that a detention pond discharges water and a retention pond does not (but rather allows the water to infiltrate).
- Retention ponds are significantly larger than detention ponds because they are required to contain all of the runoff.
- It is relatively standard for detention ponds to release flow at pre-development rates so the watershed is not receiving more than it originally could handle.

39) Most nearly, what is the shear force (kN) at support A for the simply supported beam shown?

A) 66
B) 88
C) 89
D) <u>99</u>

<u>FIND: Shear force at support A (kN)</u>

Step 1) Simplify the beam into its components:

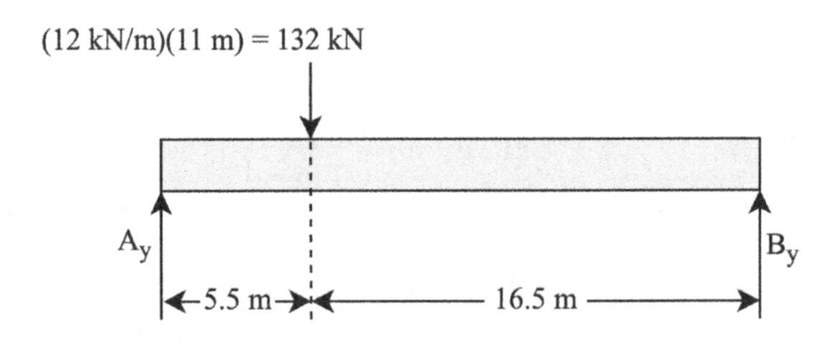

There are no horizontal forces acting on the beam. The two reaction forces are the vertical forces at supports A (R_A) and B (R_B).

Step 3) Solve for R_A by setting the moment around support B to zero:

$$M_B = \sum F_d = 0$$

$$= R_A(22\ m) - (132\ kN)(16.5\ m) = 0$$

$$R_A = \frac{(132\ kN)(16.5\ m)}{22\ m} = 99\ kN$$

$$R_A = 99\ kN$$

Notes:

- CERM Chapter 44
- The internal shear force at any given point in a beam is the resultant vertical force from both sides of the beam at that point. In this problem, the only vertical force at point A is the support reaction and so the shear force at the support is the same as the reaction.

40) A sewer pipe line has a diameter of 15 in., a slope of 0.40%, and a Manning roughness coefficient of 0.016. The uniform flow inside the pipe has a depth of 3 in. The average velocity of the flow inside the pipe (ft/s) is most nearly:

A) **<u>1.66</u>**
B) 2.34
C) 5.28
D) 16.68

Table: Depth of flow (d), Diameter (D), Area (A), and Hydraulic Radius (r_h) of a Partially Filled Circular Pipe

d/D	*area/D²*	*r_h/D*
0.19	0.1039	0.1152
0.20	0.1118	0.1206
0.21	0.1199	0.1259

FIND: Pipe velocity, v (ft/s)

Step 1) Search *manning* in the Handbook and navigate to *Chapter 6.4.5.1 Manning's Equation*:

$$v=\left(\frac{1.486}{n}\right)R_H^{\frac{2}{3}}S^{\frac{1}{2}}$$

Step 2) Determine the hydraulic radius from the given table:

$$\frac{d}{D}=\frac{3\ in}{15\ in}=0.20\ in \quad \Rightarrow\frac{r_h}{D}=0.1206=C \quad \Rightarrow r_h=CD=(0.1206)\left(\frac{15\ in}{12\frac{in}{ft}}\right)=0.1508\ ft$$

Step 3) Calculate the velocity inside the pipe:

$$v=\left(\frac{1.486}{n}\right)R_H^{\frac{2}{3}}S^{\frac{1}{2}}=\left(\frac{1.486}{0.016}\right)(0.1508\ ft)^{\frac{2}{3}}(0.004)^{\frac{1}{2}}=1.66\ \frac{ft}{s}$$

Notes:

- CERM Chapter 19
- *Steady* flow does not vary over <u>time</u> (constant-volume). *Uniform* flow does not vary over <u>space</u> (speed and pressure is equal everywhere in flow). Uniform flow simplifies problems but does not occur often. There are also *varied* and *nonuniform* flows. Know the different types of flows.
- Slope and material type are required for using the Mannings Equation!
- Doing the "decimal bounce" on paper, as opposed to in your head, to convert S from % to decimals will potentially save you errors.

<u>- END OF SOLUTIONS VERSION 1 -</u>

1) In horizontal curve elements, which of the following is most accurate regarding how deflection angles are related to corresponding arcs?

A) The deflection angle between a tangent and an arc length is a third of the arc's subtended angle.
B) The angle between two chords is a third of the arc's subtended angle.
C) The deflection angle between a tangent and a chord is half of the arc's subtended angle.
D) The angle between two chords is double the arc's subtended angle.

2) An evaporation pan with a pan coefficient of 0.80 shows a 1 day evaporation loss of 0.10 feet. Most nearly, what is the approximate evaporation loss (in.) in the adjacent fire fill pond?

A) 0.08
B) 0.75
C) 0.96
D) 8.00

3) A sand cone test is performed in the field to determine the actual field density of a construction site's compacted fill. A hole is dug in the fill and the weight of the extracted soil is measured at 1.5 kg. The whole device is filled with standard sand. The cone is placed on the hole and the sand drains to fill the hole and cone. The final weight of the device with the remaining sand is measured at 2.8 kg. The sand cone procedure is provided. The density (kg/m^3) of the compacted fill is most nearly:

Sand cone procedure:

1. Choose a location in the area to be tested. Level the ground as much as possible and dig a hole with a slightly larger diameter than the cone to approximately 8 inches deep.
2. Extract and weigh extracted soil (weight: w_1).
3. Fill the device with standard sand (~1,600 kg/m^3 or ~100 lbm/ft^3) (density: ϱ_{sand}).
4. Weight sand and device together (weight: w_2).
5. Place the device on top of the hole.
6. Open the valve and allow sand to fill the hole and cone.
7. Close valve and weigh device with remaining sand (weight: w_3).
8. Determine the weight of the sand that filled both the hole and the cone (weight: w_4=w_2-w_3).
9. Determine the volume of sand by dividing w_4 by the density of the sand (volume: V_4=w_4/ϱ_{sand}).
10. Determine the volume of the hole by subtracting the volume of the cone from the volume of the sand (volume: V_1=V_4-V_{cone})
11. Determine the density of the fill by dividing the extracted soil by the volume of the hole (ϱ_{soil}=w_1/V_1).

A) 585
B) 833
C) 851
D) 901

ϱ_{sand} = 1,600 kg/m^3
Initial weight of whole device filled with sand = 7 kg
V_{cone} = 775 cm^3

4) Which terms are used to calculate the nominal annual interest rate?

A) Nominal annual interest rate, annual effective interest rate
B) Expected life of an asset, inflation adjusted interest rate per interest period
C) Annual effective interest rate, number of compounding periods per quarter
D) Annual effective interest rate, number of compounding periods per year

5) The lap factor equation is shown below. Most nearly, how many linear feet (ft) of 7 ⅞ in. lap siding should be ordered to cover 4,000 ft² if the overlap of the siding is 2 in. and 5% waste is assumed?

A) 6,100
B) 8,100
C) 8,200
D) 8,580

$$lap\ factor = \frac{width\ of\ siding}{width\ of\ siding - overlap\ dimension}$$

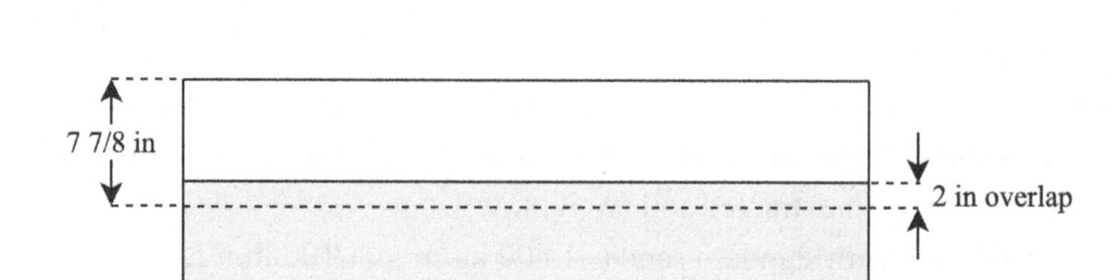

6) A construction zone is in the middle of a steep slope. What is the most appropriate run-on prevention technique?

A) Silt fence above the construction site
B) A temporary diversion swale installed above the construction site
C) Earthen berm around the entire perimeter
D) Hay bales on the downstream side of the site

7) A square footing rests on sand that has a density of 2,200 kg/m³ and an angle of internal friction of 30°. The shape factor, *s*, for a square footing is 0.85. The equation for the net bearing capacity for sand is provided. Using the Terzaghi bearing capacity factors, what is most nearly the width of the footing (m) if it is to be placed 1 m below the surface and the net design capacity is 3302/B²?

$$q_{net} = \frac{1}{2} B\rho g N_{\gamma} s + \rho g D_f \left(N_q - 1 \right)$$

A) 0.5
B) 1.0
C) 1.5
D) 2.0

8) A 10 ft x 10 ft footing exerts a pressure of 4,300 psf on the soil below it. Most nearly, what is the increase in vertical soil stress (psf) 20 ft below and 20 ft from the center of the footing?

A) 82
B) 84
C) 86
D) 88

9) Shear stress can be induced in a beam due to bending. Which of the follow-up statements is most likely false?

A) Horizontal shear exists even when the loading is vertical.
B) The bending moment is positive if it produces bending of the beam concave downwards (compression in bottom fibers and tension in top fibers).
C) One set of parallel shears counteracts the rotational moment from the other set of parallel shears.
D) In biaxial loading, identical shear stresses exist simultaneously in all four directions.

10) Based on the following traffic counts, the peak hour factor is most nearly:

A) 0.82
B) 0.81
C) 1.16
D) 1.0

Time Interval	Volume (vehicles)
1:15-1:30	634
1:30-1:45	456
1:45-2:00	346
2:00-2:15	478
2:15-2:30	565
2:30-2:45	498
2:45-3:00	765
3:00-3:15	654
Total	**4,396**

11) A possible reason for Hydraulic Grade Line below the pipe invert elevation is:

A) Vapor pressure
B) High head loss
C) Low velocity head
D) Negative pressure

12) At 86°F, two untied 8 feet wide sheets of copper abut each other. The coefficient of linear thermal expansion of copper is 8.9×10^{-6} degree F^{-1}. What is their maximum possible separation (in.) at 42°F?

A) 0.028
B) 0.075
C) 0.122
D) 1.000

13) A concrete masonry unit (CMU) wall is reinforced in the center and partially grouted. The wall has no. 8 vertical bars spaced 40 in. apart, which provides a reinforcing steel area, A, of 0.236 in^2/ft. Using the equations provided, what is most nearly the ratio of the distance between the compression face of the wall and neutral axis to the effective depth, k?

A) 0.003
B) 0.05214
C) 0.263
D) 0.236

Actual wall thickness, $b = 12$ inches
Modular ratio, $n = 15.8$
d = depth to reinforcement
ϱ = ratio of tensile steel area to gross area of masonry

$$k = \sqrt{2\rho n + (\rho n)^2} - \rho n$$

$$\rho = \frac{A_s}{bd}$$

14) A job status report shows the following information about a $275,000 project. The annual cost of the work performed is $125,000. As of today, 90% of the project should be completed but only 75% has actually been completed. The schedule variance ($) for the project is most nearly:

A) -206,250
B) -41,250
C) 40,250
D) 247,500

15) A surveyor sets up his instrument first at point A and then at point B, and reads elevations with her rod at points 1, 2, and 3. Point 1 is 4769 ft above mean sea level. Backsight (BS) and foresight (FS) information is shown. The ground elevation (ft) at point 3 is most nearly:

A) 4759.27
B) 4759.32
C) 4763.70
D) 4769.00

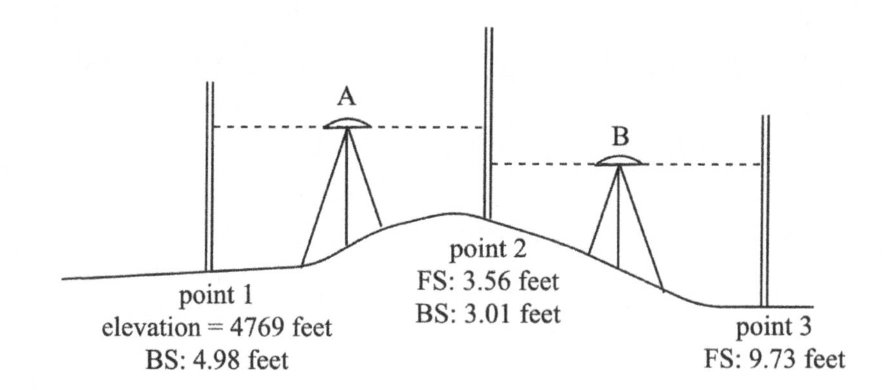

16) Which four are the most likely reasons for foundation settlement?

A) Frost heave, soil type, varying foundation depths, water leaks
B) Frost heave, increased active loads on the first floor, varying foundation depths, vegetation
C) Soil type, decreased passive loads on the first floor, extreme weather events, frost heave
D) Heavy rainfall, soil type, vegetation, increased active loads on the second floor.

17) What is the coating area (ft^2/ft) of both sides of an NZ steel sheet pile if the wall thickness is 0.394 inches and the moment of inertia is 292.8 in^4/ft?

A) 6.18
B) 6.20
C) 6.49
D) 6.50

18) A trapezoidal channel has been designed for optimum efficiency and without freeboard, with a flow depth of 4.75 feet. The channel flow (cfs) is most nearly:

n = 0.011
Slope = 4%

A) 188
B) 376
C) 1,882
D) 2,513

19) Storm inlets must be placed at the low point of a road. The beginning (PVC) and end (PVT) stations of the curve, and grades of the road are given. Most nearly, at what station must the storm inlets be placed?

A) 29+67
B) 48+50
C) 49+15
D) 96+43

$STA_{BVC} = 25+50$
$STA_{EVC} = 96+44$
$G_1 = -1.20\%$
$G_2 = 2.40\%$

20) Which types of top driven piles have a maximum allowable driving stress that is equal to 90% of the yield stress of steel?

A) Steel H-piles, unfilled steel pipe piles, concrete-filled steel pipe piles
B) Steel H-piles, unfilled steel pipe piles, precast prestressed concrete pile
C) Timber piles, precast prestressed concrete pile, conventionally reinforced concrete piles
D) Steel L-piles, filled steel pipe piles, timber piles

21) The foundation excavation shown has an 11-ft deep braced cut in stiff clay. If the wall is supporte
during construction by a temporary structure, the total earth force per foot of wall (lbf/ft) is most nearly

A) 2,386.2
B) 3,112.5
C) 3,838.7
D) 4,565.0

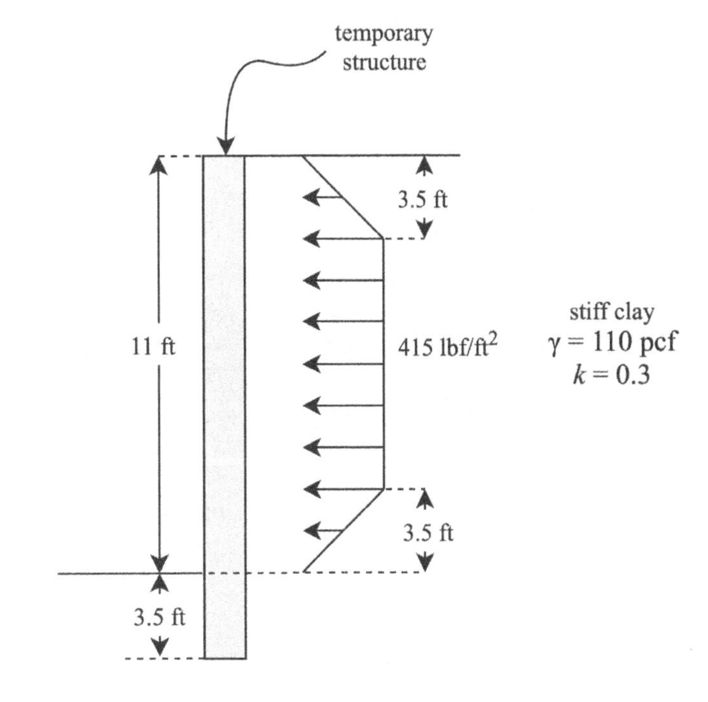

22) A crew is able to place 35 yd^3 of concrete per 8-hr work day. The project requires 2,098 ft^3 of concrete to be placed. Most nearly, what is the total cost ($) of the personnel involved in concreting if the crew consists of one foreman at $65/hr and three laborers at $32/hr?

A) 2,737
B) 2,860
C) 4,364
D) 8,578

dation is being constructed that is 95 ft x 100 ft. Soil must be replaced to a depth
the foundation. The soil will be borrowed from a nearby pit. Formulas are given
oil factors. The loose volume of soil (yd^3) that is needed to be transported from
the new foundation is most nearly:

A) 2,674
B) 2,971
C) 3,118
D) 3,120

γ_{borrow} = 95 pcf
Swell = 5%
Shrinkage = 10%

Quantity	Symbol	Units	Formulas		
bank volume	BCY	yd^3	$\frac{LCY}{swell\,factor}$	$\frac{CCY}{(1-shrinkage)}$	
loose volume	LCY	yd^3	*Shrinkage factor x BCY*	$\frac{BCY}{load\,factor}$	$\frac{\gamma_{bank}}{\gamma_{loose}} x\,BCY$

24) An 11 ft cut in sandy soil is being supported by a reinforced concrete wall. The backfill is level and a surcharge pressure of 500 lbf/ft^2 extends behind the wall. Information about the soil is given and a diagram is shown below. If the surcharge resultant acts halfway up from the base and the active soil resultant acts one-third up from the base, the total overturning moment (lbf) per foot of wall, taken about the toe, is most nearly:

A) 5,765
B) 56, 211
C) 72,787
D) 145,573

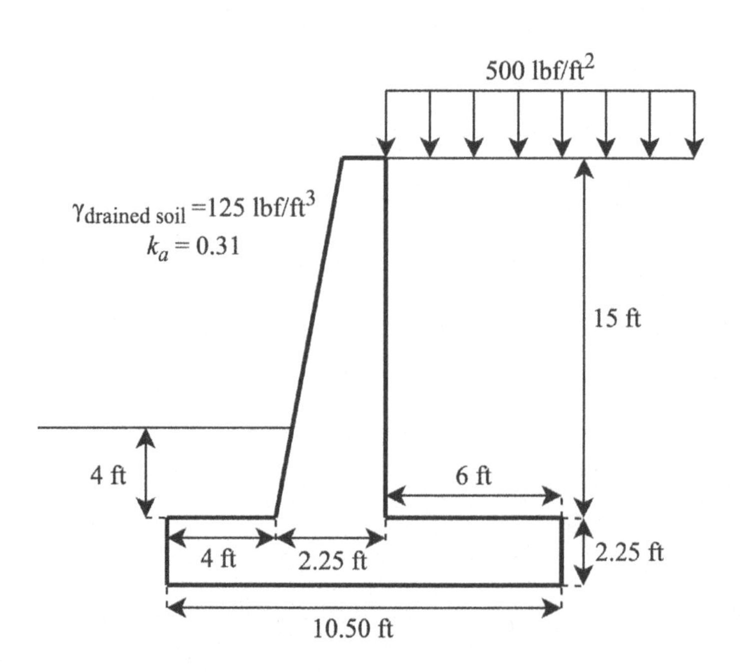

25) Which one of the following is most accurate regarding steel column design?

A) As a column becomes longer, the load that causes the buckling becomes smaller.
B) As a column becomes wider, the load that causes the buckling becomes smaller.
C) As a column becomes longer, the load that causes the buckling becomes larger.
D) As a column becomes shorter, the load that causes the buckling becomes smaller.

26) Which of the following equations describes the OSHA Incidence Rate?

A) (Number of injuries, illnesses, and fatalities) x (200,000) / (Total hours worked by all employees during the period in question)
B) (Total number of recordable incidents resulting in days away, restricted, or transferred) x (200,000) / (Total hours worked by all employees)
C) (Total number of lost time incidents) x (200,000) / (Total number of employees worked each year)
D) (Total number of recordable incidents) x (200,000) / (Total number of hours worked by injured employees)

27) Water is carried through a 311 ft long, 8 in. (internal diameter) schedule 40 welded and seamless pipe ($\mathcal{E}$=0.0002 ft) that contains one fully open flanged steel angle valve, two flanged steel regular 90° elbows, and one flanged steel gate valve. The discharge is located 20 ft higher than the intake. Taking into consideration minor losses, the total head loss (ft) through the pipe is most nearly:

A) 361.70
B) 371.90
C) 381.70
D) 392.00

$Re = 1.92 \times 10^5$

$v = 45$ ft/s

28) Which of the following is the most appropriate friction factor to use in calculating earth pressure for a foundation wall interfacing with stiff clay?

A) 0.32
B) 0.70
C) 0.22
D) 1.10

29) Soil was extracted from a depth of 7 m. The grain size distribution characteristics, and liquid and plastic limits of the soil are as shown below. What is the classification of the soil according to the USCS?

A) SP
B) SW
C) CH
D) SC

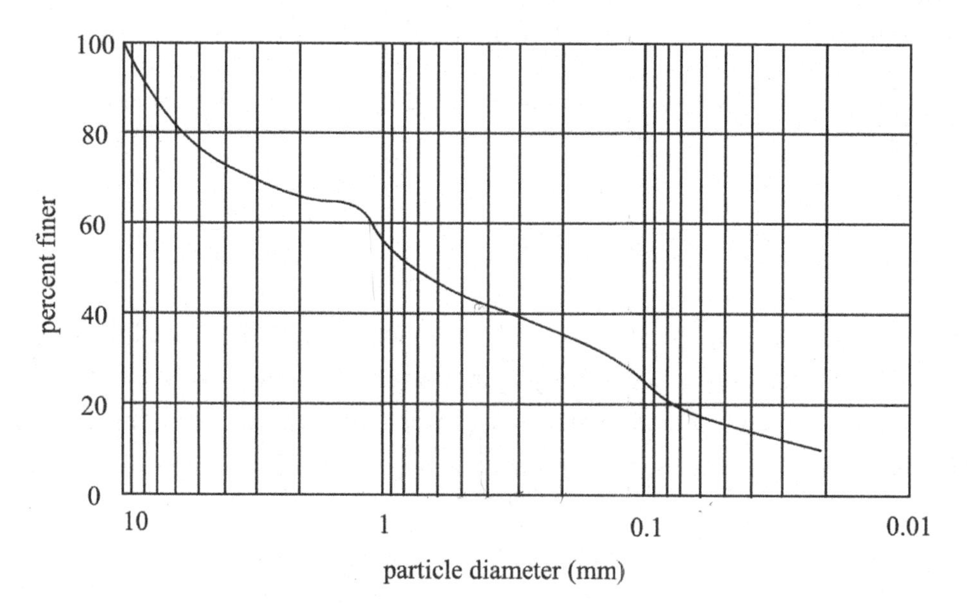

LL = 33.1
PL = 10.9

30) The layout of a wall is shown. Allow for 4% brick waste. Mortar joints are required to be 0.5 in. thick. Two rows of brick positioned as stretchers are required. Using nominal sizing, most nearly, how many standard size non-modular bricks are needed?

A) 369
B) 370
C) 739
D) 769

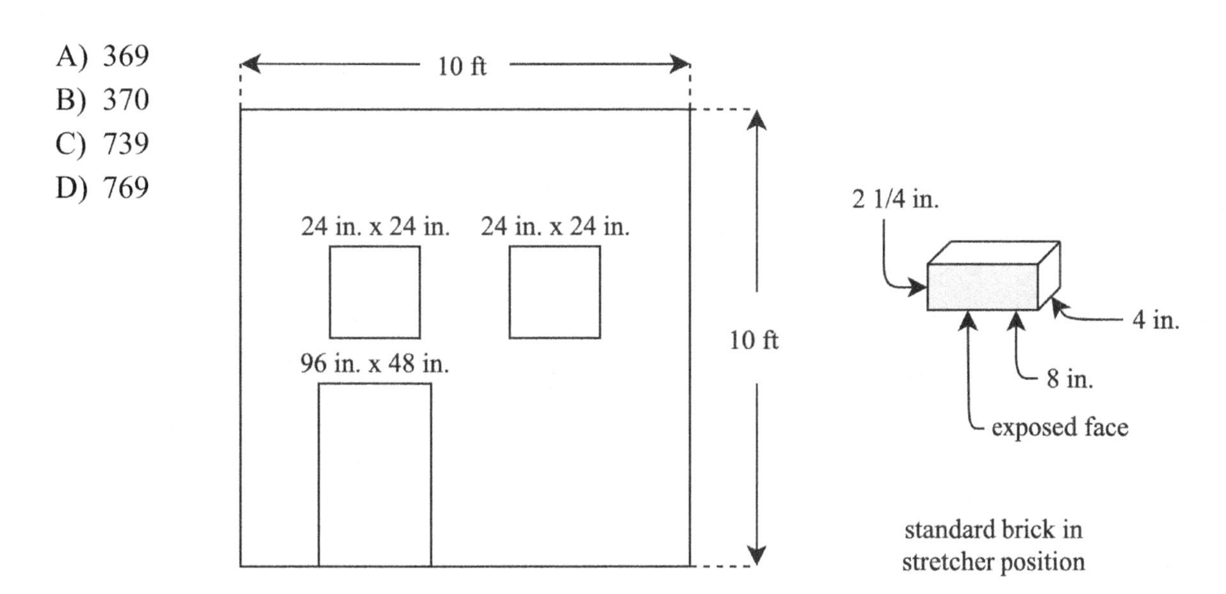

31) What is the slope of the initial linear portion of the curve on the stress-strain diagram for steel?

A) Toughness
B) Modulus of Elasticity
C) Yield stress
D) Shear modulus

32) A 2-hr storm over a 95 km^2 watershed area produces a total runoff volume of $4.5x10^6$ m^3 with a peak discharge of 325 m^3/s. If a 2-hr storm producing 8 cm of runoff is to be used to design a culvert in the same area, the design discharge (m^3/s) is most nearly:

A) 55
B) 69
C) 324
D) 553

33) The saturated unit weight (kN/m^3) of a collected soil sample with the given characteristics is most nearly:

A) 17.66
B) 17.76
C) 19.58
D) 19.68

SG = 2.67
$e_{max} = 0.92$
$e_{min} = 0.34$
$D_r = 44\%$

34) It takes 7 minutes and 45 seconds for a 30 mm thick two-way drained clay layer to consolidate 30% in the lab. Most nearly, how many days are required for a 9 m thick two-way drained layer of the same clay to reach 30% consolidation in the field?

A) 8
B) 484
C) 501
D) 1,938

35) An 86-ac watershed area is shown. Rain gauge stations and their recorded rainfall values are marked. The isohyets have also been drawn. The approximate mean precipitation (mm) of the catchment area is most nearly:

A) 2.79
B) 3.07
C) 5.30
D) 6.88

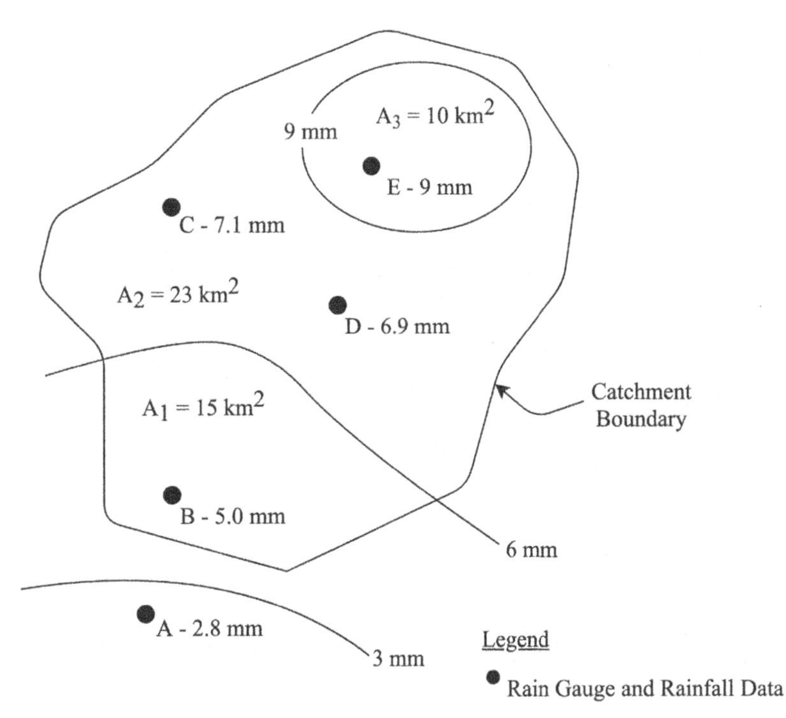

36) Which of the following is the most accurate description of moderately weathered rock?

A) Rock shows slight discoloration.
B) More than half of the rock is decomposed.
C) Less than half of the rock is decomposed.
D) Original minerals of rock have almost entirely decomposed.

37) The section modulus, S, of a rectangular b x h section is calculated as: $S_{rectangular} = bh^2/6$. The maximum bending stress at midspan (kN/m^2) for the simply supported beam shown below is most nearly:

A) 48
B) 64
C) 128
D) 192

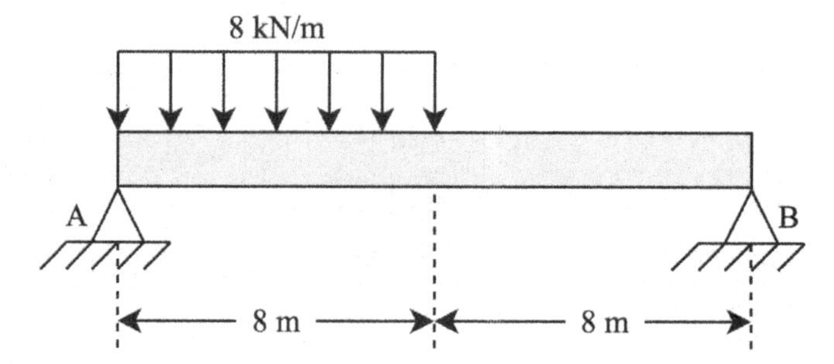

38) A 2:1 (horizontal:vertical) sloped cut is made in submerged clay. The cut is 38 feet deep and the clay extends 19 ft below the toe of the cut before it hits a rock layer. The cohesive factor of safety of the cut is 1.5. The saturated density of the clay (lbf/ft^3) is most nearly:

A) 54
B) 132
C) 178
D) 194

$c = 1{,}200$ lbf/ft^2
$F_{cohesive} = 1.5$

39) A road is required to have a 3% crown to properly drain stormwater runoff to the gutters. The gutter detail is shown below. If the centerline of the road is 4783.14 ft and the road is 34 ft wide (TBC to TBC), the elevation of the flow line (ft) is most nearly:

A) 4781.38
B) 4782.61
C) 4782.63
D) 4782.69

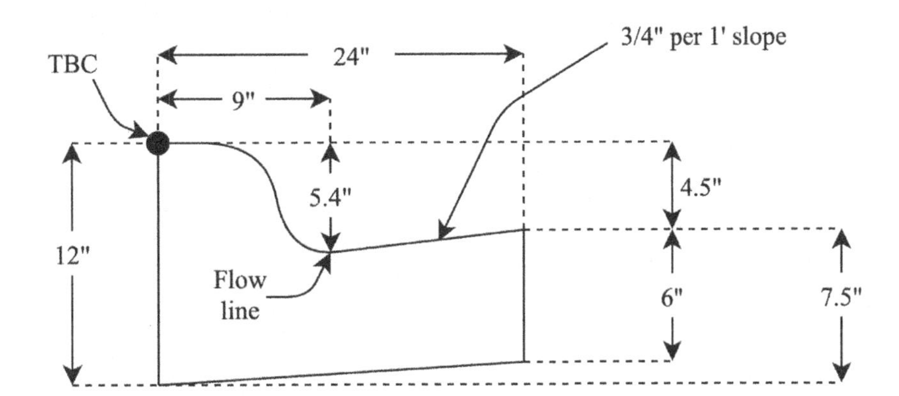

40) The structure shown below carries a dead load of 45 psf (including self weight) and a live load of 55 psf. The unfactored axial stress (ksi) in the center column is most nearly:

A) 0.25
B) 0.28
C) 1.50
D) 1.55

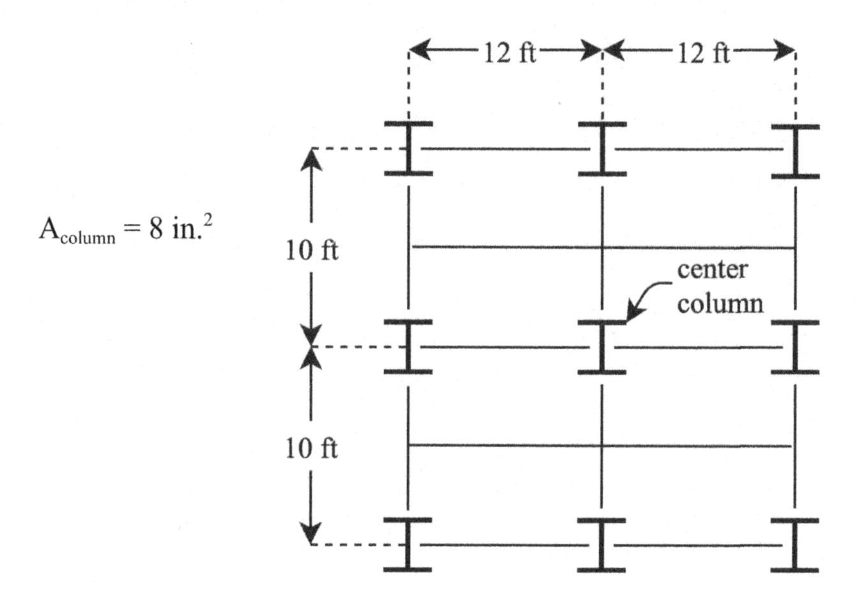

- END OF EXAM VERSION 2 -

1) In horizontal curve elements, which of the following is most accurate regarding how deflection angles are related to corresponding arcs?

A) The deflection angle between a tangent and an arc length is a third of the arc's subtended angle.
B) The angle between two chords is a third of the arc's subtended angle.
C) The deflection angle between a tangent and a chord is half of the arc's subtended angle.
D) The angle between two chords is double the arc's subtended angle.

Search *horizontal curve* and navigate to *Chapter 5.2.1 Basic Curve Elements* to deduce the most accurate answer from the Parts of a Circular Curve diagram.

Notes: CERM Chapter 79

2) An evaporation pan with a pan coefficient of 0.80 shows a 1 day evaporation loss of 0.10 ft. Most nearly, what is the approximate evaporation loss (in.) in the adjacent fire fill pond?

A) 0.08
B) 0.75
C) 0.96
D) 8.00

FIND: Reservoir Evaporation, E_R (inches)

Step 1) Search *evaporation* in the Handbook and navigate to the Pan Method in *Chapter 6.5.8.3 Evaporation*:

$$E_L = K_p E_p$$

$$= (0.80)(0.10\,ft)\left(\frac{12\ in}{ft}\right) = 0.96\,ft$$

Notes:

- CERM Chapter 20
- The beginning of each chapter lists the symbol, subscript, and unit of each variable. Take advantage of this when you feel lost!

3) A sand cone test is performed in the field to determine the actual field density of a construction site's compacted fill. A hole is dug in the fill and the weight of the extracted soil is measured at 1.5 kg. The whole device is filled with standard sand. The cone is placed on the hole and the sand drains to fill the hole and cone. The final weight of the device with the remaining sand is measured at 2.8 kg. The sand cone procedure is provided. The density (kg/m^3) of the compacted fill is most nearly:

Sand cone procedure:

1. Choose a location in the area to be tested. Level the ground as much as possible and dig a hole with a slightly larger diameter than the cone to approximately 8 inches deep.
2. Extract and weigh extracted soil (weight: w_1).
3. Fill the device with standard sand (~1,600 kg/m^3 or ~100 lbm/ft^3) (density: ϱ_{sand}).
4. Weight sand and device together (weight: w_2).
5. Place the device on top of the hole.
6. Open the valve and allow sand to fill the hole and cone.
7. Close valve and weigh device with remaining sand (weight: w_3).
8. Determine the weight of the sand that filled both the hole and the cone (weight: $w_4=w_2-w_3$).
9. Determine the volume of sand by dividing w_4 by the density of the sand (volume: $V_4=w_4/\varrho_{sand}$).
10. Determine the volume of the hole by subtracting the volume of the cone from the volume of the sand (volume: $V_1=V_4-V_{cone}$)
11. Determine the density of the fill by dividing the extracted soil by the volume of the hole ($\varrho_{soil}=w_1/V_1$).

A) 585
B) <u>833</u>
C) 851
D) 901

ϱ_{sand} = 1,600 kg/m^3
Initial weight of whole device filled with sand = 7 kg
V_{cone} = 775 cm^3

FIND: Density of compacted fill (kg/m^3)

Based on the procedure given, the calculations are as follows:

Mass of sand in hole and cone: $w_4=w_2-w_3 = 4.2\ kg$

Volume of sand in hole and cone: $V_4 = w_4 / \varrho_{sand} = 4.2\ kg / 1{,}600\ kg/m^3 = 0.0026\ m^3$

Volume of hole: $V_1 = V_4 - V_{cone} = 0.0026\,m^3 - (775\,cm^3)(m/100\,cm)^3 = 0.0018\,m^3$

Density of compacted fill: $\varrho_{sand} = w_1 / V_1 = 1.5\,kg/0.0018\,m^3 = 833\,kg/m^3$

4) Which terms are used to calculate the nominal annual interest rate?

A) Nominal annual interest rate, annual effective interest rate
B) Expected life of an asset, inflation adjusted interest rate per interest period
C) Annual effective interest rate, number of compounding periods per quarter
D) Annual effective interest rate, number of compounding periods per year

Search *nominal annual interest rate* and navigate to *Chapter 1.7.1 Nomenclature and Definitions.* Rearrange the equation for Non-annual Compounding to solve for the nominal annual interest rate.

5) The lap factor equation is shown below. Most nearly, how many linear feet (ft) of 7 ⅞ in. lap siding should be ordered to cover 4,000 ft^2 if the overlap of the siding is 2 in. and 5% waste is assumed?

A) 6,100
B) 8,100
C) 8,200
D) 8,580

$$lap\ factor = \frac{width\ of\ siding}{width\ of\ siding - overlap\ dimension}$$

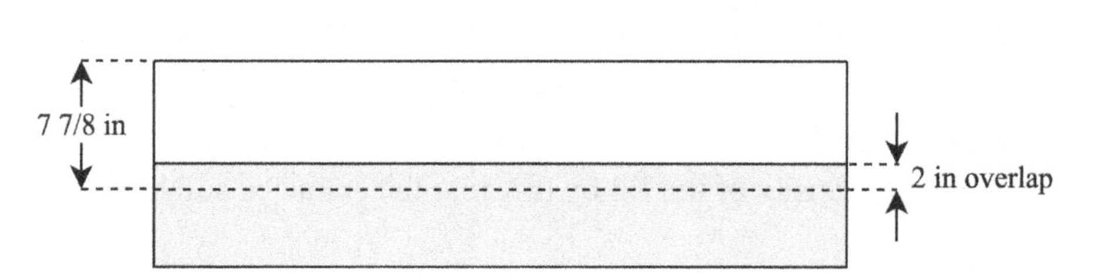

FIND: Linear feet of lap siding (ft)

Step 1) Calculate the lap factor using the given equation:

$$lap\ factor = \frac{width\ of\ siding}{width\ of\ siding - overlap\ dimension} = \frac{7.875\ in}{7.875\ in - 2\ in} = 1.34$$

Step 2) Multiply the area to be covered by the lap factor and by the assumed waste:

$$(4{,}000\ ft^2)(1.34)(1.05) = 5{,}628\ ft^2$$

Step 3) Calculate the number of linear feet by dividing the area by the width of the siding pieces:

$$\frac{(5{,}628\ ft^2)\left(12\ \frac{in}{ft}\right)}{7.875\ in} = 8{,}576\ ft$$

Notes:

- Always round up on quantities, otherwise you'll come up short!

6) A construction zone is in the middle of a steep slope. What is the most appropriate run-on prevention technique?

A) Silt fence above the construction site
B) A temporary diversion swale installed above the construction site
C) Earthen berm around the entire perimeter
D) Hay bales on the downstream side of the site

A temporary diversion swale installed above the construction site is most appropriate to control run-on.

Silt fences should only be placed at the bottom of slopes. Earthen berms should only be placed at the bottom of shallow slopes. Hay bales on the downstream side of the site would not prevent run-on.

7) A square footing rests on sand that has a density of 2,200 kg/m^3 and an angle of internal friction of 30°. The shape factor, *s*, for a square footing is 0.85. The equation for the net bearing capacity for sand is provided. Using the Terzaghi bearing capacity factors, what is most nearly the width of the footing (m) if it is to be placed 1 m below the surface and the net design capacity is 3302/B^2?

$$q_{net} = \frac{1}{2} B\rho g N_\gamma s + \rho g D_f \left(N_q - 1 \right)$$

A) 0.5
B) 1.0
C) 1.5
D) <u>2.0</u>

FIND: Width of footing, B (m)

Step 1) Search *bearing capacity* and navigate to *Chapter 3.4.2.1 Bearing Capacity Equation for Concentrically Loaded Square or Rectangular Footings.*

Step 3) Determine from the Bearing Capacity Factors table that N_q=18.4 and N_γ=22.4.

Step 4) Calculate the net bearing capacity:

$$q_{net} = \frac{1}{2} B\rho g N_\gamma s + \rho g D_f \left(N_q - 1 \right)$$

$$= \frac{\frac{1}{2} B\left(2,200 \frac{kg}{m^3}\right)\left(9.81 \frac{m}{s^2}\right)(22.4)(0.85)}{1,000 \frac{Pa}{kPa}} + \frac{\left(2,200 \frac{kg}{m^3}\right)\left(9.81 \frac{m}{s^2}\right)(1\ m)(18.4 - 1)}{1,000 \frac{Pa}{kPa}}$$

$$= 205.46 B\ kPa + 375.53\ kPa$$

Step 5) Equate the two expressions for q_{net} and calculate B:

$$\frac{3302}{B^2} = 205.46 B\ kPa + 375.53\ kPa$$

$$B = 2\ m$$

Notes:

- CERM Chapter 36
- Be prepared to interpolate for bearing capacity and shape factors if necessary.
- Use your calculator's solver or substitute in the given answers to speed up the last step!

8) A 10 ft x 10 ft footing exerts a pressure of 4,300 psf on the soil below it. Most nearly, what is the increase in vertical soil stress (psf) 20 ft below and 20 ft from the center of the footing?

A) 82
B) 84
C) <u>86</u>
D) 88

FIND: Increase in vertical soil stress, ΔP (psf)

Step 1) Search *square footing* in the Handbook and navigate to the Boussinesq contour chart in *Chapter 3.5.1 Stress Distribution*. The equation for this chart is:

$$\Delta p_v = I(no.\ of\ \ squares)\, p_{app}$$

Step 2) Determine from the *Boussinesq contour chart* that the line of influence for a square foundation at (2B, -2B) is 0.02p.

Step 3) Calculate the increase in vertical soil stress:

$$\Delta p = (0.02)\,(4,300\ pcf) = 86\ pcf$$

Notes:

- CERM Chapter 40
- Don't feel bad if you find the applied load pressure graphs a bit confusing! In this problem, the width of the footing, B, is 10 ft. When you look at the Boussinesq chart, you are looking for the increase in vertical soil stress at the point specified - so 20 ft below and 20 ft from the center of the footing. Therefore, you go over 2B (20 ft = 2x10 ft) and down 2B, which are the "squares" referred to in the equation.
- An aside, remember that an "infinitely long footing" is simply one that is not a square.

9) Shear stress can be induced in a beam due to bending. Which of the follow-up statements is most likely false?

A) Horizontal shear exists even when the loading is vertical.
B) <u>The bending moment is positive if it produces bending of the beam concave downwards (compression in bottom fibers and tension in top fibers).</u>
C) One set of parallel shears counteracts the rotational moment from the other set of parallel shears.
D) In biaxial loading, identical shear stresses exist simultaneously in all four directions.

Search shear stress and navigate to *Chapter 1.6.7.1 Shearing Force and Bending Moment Sign Conventions*.

Notes: CERM Chapter 44

10) Based on the following traffic counts, the peak hour factor is most nearly:

A) 0.82
B) 0.81
C) 1.16
D) 1.0

Time Interval	Volume (vehicles)
1:15-1:30	634
1:30-1:45	456
1:45-2:00	346
2:00-2:15	478
2:15-2:30	565
2:30-2:45	498
2:45-3:00	765
3:00-3:15	654
Total	**4,396**

FIND: Peak hour factor (PHF)

Step 1) Search *peak hour factor* in the Handbook and navigate to *Chapter 5.1.3.3 Peak-Hour Factor*:

$$PHF = \frac{V}{V_{15} \cdot 4}$$

Step 2) Calculate the hourly traffic volume for each interval to find the maximum actual hourly volume:

Time Interval	Hourly Volume (vehicles)
1:15-2:15	1,914
1:30-2:30	1,845
1:45-2:45	1,887
2:00-3:00	2,306
2:15-3:15	2,482

Step 3) Calculate the peak hour factor:

$$PHF = \frac{V}{V_{15} \cdot 4} = \frac{2,482 \frac{veh}{hr}}{\left(765 \frac{veh}{period}\right)\left(4 \frac{periods}{hr}\right)} = 0.81$$

Notes: CERM Chapter 73

11) A possible reason for Hydraulic Grade Line below the pipe invert elevation is:

A) Vapor pressure
B) High head loss
C) Low velocity head
D) Negative pressure

The *hydraulic grade line* (HDL) is defined in *Chapter 6.2.1.4 Hydraulic Gradient* as the line connecting the sum of pressure and elevation heads at different points in conveyance systems: HGL=$h_p + h_z$.

Notes: CERM Chapter 16

12) At 86°F, two untied 8 feet wide sheets of copper abut each other. The coefficient of linear thermal expansion of copper is $8.9\text{x}10^{-6}$ degree F^{-1}. What is their maximum possible separation (in.) at 42°F?

A) 0.028
B) 0.075
C) 0.122
D) 1.000

FIND: Maximum separation (inches)

Step 1) Search *thermal expansion* in the Handbook and navigate to *Chapter 1.6.3 Thermal Deformations:*

$$\delta_t = \alpha L\left(T - T_0\right)$$
$$= \left(8.9x10^{-6}\frac{1}{^\circ F}\right)(8\,ft)\left(12\frac{in}{ft}\right)(86^\circ F - 42^\circ F)$$
$$= 0.0376\ in$$

Step 2) Calculate the total separation for two slabs:

$$Total\ separation = 2\ slabs\ x\ 0.0376\ in/slab = 0.0752\ in$$

Notes:

- CERM Chapter 44
- The problem itself does not really have a good keyword for beginning to search for a solution to this. However, the word "thermal" is automatically a keyword to start with because it involves temperature.

13) A concrete masonry unit (CMU) wall is reinforced in the center and partially grouted. The wall has no. 8 vertical bars spaced 40 in. apart, which provides a reinforcing steel area, A, of 0.236 in^2/ft. Using the equations provided, what is most nearly the ratio of the distance between the compression face of the wall and neutral axis to the effective depth, k?

A) 0.003
B) 0.05214
C) <u>0.263</u>
D) 0.236

Actual wall thickness, b = 12 inches
Modular ratio, n = 15.8
d = depth to reinforcement
ϱ = ratio of tensile steel area to gross area of masonry

$$k = \sqrt{2\rho n + (\rho n)^2} - \rho n$$

$$\rho = \frac{A_s}{bd}$$

FIND: Ratio of the distance between the compression face of the wall and neutral axis to the effective depth, *k*

Step 1) Reinforcement is placed in the center of the wall, so $d = 12\ in/2 = 6\ in$. Calculate ϱ using the equation provided:

$$\rho = \frac{A_s}{bd} = \frac{0.236\ in^2}{(12\ in)(6\ in)} = 0.003$$

Step 2) Calculate k:

$$\rho n = (0.003)(15.8) = 0.047$$

$$k = \sqrt{2\rho n + (\rho n)^2} - \rho n = \sqrt{(2)(0.047) + (0.047)^2} - 0.047 = 0.263$$

Notes: CERM Chapter 68

14) A job status report shows the following information about a $275,000 project. The annual cost of the work performed is $125,000. As of today, 90% of the project should be completed but only 75% has actually been completed. The schedule variance ($) for the project is most nearly:

A) -206,250
B) -41,250
C) 40,250
D) 247,500

Find: Schedule variance, SV ($)

Step 1) Search *schedule variance* in the Handbook and navigate to *Chapter 2.4.1.3 Earned-Value Analysis:*

$$SV = BCWP - BCWS$$

Step 2) The Budgeted Cost of Work Performed, BCWP, is the actual earned value based on what work has actually been accomplished. According to the CERM, it can be calculated by multiplying the fraction of work completed by the planned cost of the project:

$$BCWP = \$275{,}000 \times 75\% = \$206{,}250$$

Step 3) The Budgeted Cost of Work Scheduled, BCWS, is calculated by multiplying the total budget by the percentage of work that should be completed to-date:

$$BCWS = \$275{,}000 \times 90\% = \$247{,}500$$

Step 4) Calculate the schedule variance:

$$SV = BCWP - BCWS = \$206{,}250 - \$247{,}500 = -\$41{,}250$$

Notes:

- CERM Chapter 86
- Understanding the various components of the EVM may be useful!

15) A surveyor sets up her instrument first at point A and then at point B, and reads elevations with her rod at points 1, 2, and 3. Point 1 is 4769 ft above mean sea level. Backsight (BS) and foresight (FS) information is shown. The ground elevation (ft) at point 3 is most nearly:

A) 4759.27
B) 4759.32
C) <u>4763.70</u>
D) 4769.00

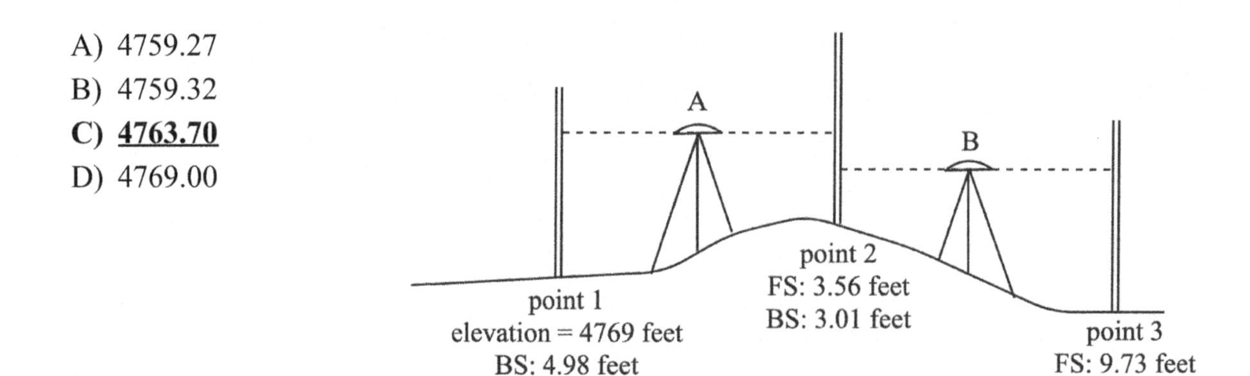

FIND: Ground elevation at point 3 (ft)

Step 1) Search *leveling* in the Handbook and navigate to *Chapter 2.1.3 Site Layout and Control.* Find the elevation difference between the point 1 BS and the point 2 FS based on the rod readings when the instrument is set up at point A:

$$elevation_{1-2} = BS_1 - FS_2 = 4.98\,ft - 3.56\,ft$$
$$= 1.42\,ft$$

Step 2) Find the elevation difference between the point 2 BS and the point 3 FS based on the rod readings when the instrument is set up at point B:

$$elevation_{2-3} = BS_2 - FS_3 = 3.01\,ft - 9.73\,ft$$
$$= -6.72\,ft$$

Step 3) Now find the elevation difference between points 1 and 3:

$$elevation_{1-3} = elevation_{1-2} + elevation_{2-3} = 1.42\,ft + (-6.72\,ft)$$
$$= -5.30\,ft$$

The elevation at point 1 is 4769 feet. Therefore, the elevation at point 3 is:

$$elevation_3 = 4769\,ft + (-5.30\,ft) = 4763.70\,ft$$

Notes:

- CERM Chapter 78
- Positive values indicate the station is higher in elevation than the previous station, and negative values indicate that the station is lower in elevation than the previous station.

16) Which four are the most likely reasons for foundation settlement?

A) **Frost heave, soil type, varying foundation depths, water leaks**
B) Frost heave, increased active loads on the first floor, varying foundation depths, vegetation
C) Soil type, decreased passive loads on the first floor, extreme weather events, frost heave
D) Heavy rainfall, soil type, vegetation, increased active loads on the second floor.

The first answer is the only common sense answer. Vegetation would not likely cause foundation settlement; foundations should be designed for all active and passive loads on all floors, and; decreased loads would not likely ever cause settlement.

17) What is the coating area (ft^2/ft) of both sides of an NZ steel sheet pile if the wall thickness is 0.394 inches and the moment of inertia is 292.8 in^4/ft?

A) **6.18**
B) 6.20
C) 6.49
D) 6.50

Search *sheet pile* and navigate to the table in *Chapter 4.2.2 Steel Sheet Pile Properties.*

18) A trapezoidal channel has been designed for optimum efficiency and without freeboard, with a flow depth of 4.75 feet. The channel flow (cfs) is most nearly:

A) 188
B) 376
C) 1,882
D) 2,513

n = 0.011
Slope = 4%

FIND: Flow (cfs)

Step 1) Search *manning's equation* in the Handbook and navigate to *Chapter 6.4.5.1 Manning's Equation* for the equation for flow, Q:

$$Q=\frac{1.486}{n}AR_H^{\frac{2}{3}}S^{\frac{1}{2}}$$

Step 2) Search *efficient section* and navigate to the table entitled Best Hydraulic Efficient Sections Without Freeboard in *Chapter 6.4.5.4 Flow in Channels*. Calculate Area, A:

$$A=\sqrt{3}y^2=\sqrt{3}\cdot(4.75\,ft)^2=39.08\,ft^2$$

Step 3) Calculate the Hydraulic Radius, R:

$$R=\frac{1}{2}y=\frac{1}{2}\cdot 4.75\,ft=2.38\,ft$$

Step 4) Plug A and R into the equation for flow:

$$Q=\frac{1.486}{n}AR_H^{\frac{2}{3}}S^{\frac{1}{2}}=\frac{1.486}{0.011}(39.08\,ft^2)(2.38\,ft)^{\frac{2}{3}}(0.04)^{\frac{1}{2}}=1,882.18\frac{ft^3}{s}$$

Notes: CERM Chapter 19

19) Storm inlets must be placed at the low point of a road. The beginning (PVC) and end (PVT) stations of the curve, and grades of the road are given. Most nearly, at what station must the storm inlets be placed?

A) 29+67
B) 48+50
C) 49+15
D) 96+43

$STA_{BVC} = 25+50$
$STA_{EVC} = 96+44$
$G_1 = -1.20\%$
$G_2 = 2.40\%$

FIND: Station of low point

Step 1) Search *vertical curves* in the Handbook and navigate to *Chapter 5.3.1 Symmetrical Vertical Curve Formula.* Find the equation for the horizontal distance to min/max elevation on curve, x_m:

$$x_m = \frac{g_1 L}{g_1 - g_2}$$

Step 2) Calculate the length of the curve, L:

$$L = (Sta.\ 96+44) - (Sta.25+50) = 70.94\ stations$$

Step 3) Calculate the turning point location in relation to the station of the PVC:

$$x_m = \frac{g_1 L}{g_1 - g_2} = \frac{(-1.20\%)(70.94\ sta.)}{-1.20\% - 2.40\%} = 23.66\ stations \approx 24\ stations$$

Step 4) Determine the station of the low/minimum point:

$$Sta._x = Sta._{BVC} + x_m = (Sta.\ 25+50) + 24\ stations = Sta.\ 49+17$$

Notes:

- CERM Chapter 79
- Remember that Station 49+50.00 is 4950.00 feet past Station 00+00.00.
- The distance between Sta. 96+44 and Sta. 25+50 is 7,094 feet, or 70.94 stations, since stations are distances of 100 feet.
- Remember that for vertical curves, PI is located at L/2.
- Note that g is a decimal and L is length of curve in feet. You can either use those units, or you can keep g in percent and make L in stations as is shown in this problem. Be careful not to mix and match though!
- This enumeration of stationing is used regularly in design and construction of lineal infrastructure such as roads, sewer systems, water systems, etc.

20) Which types of top driven piles have a maximum allowable driving stress that is equal to 90% of the yield stress of steel?

A) <u>Steel H-piles, unfilled steel pipe piles, concrete-filled steel pipe piles</u>
B) Steel H-piles, unfilled steel pipe piles, precast prestressed concrete pile
C) Timber piles, precast prestressed concrete pile, conventionally reinforced concrete piles
D) Steel L-piles, filled steel pipe piles, timber piles

Navigate to the table in the Handbook's Chapter 2 entitled Allowable Stresses in Pile for Top-Driven Piles.

21) The foundation excavation shown has an 11-ft deep braced cut in stiff clay. If the wall is supported during construction by a temporary structure, the total earth force per foot of wall (lbf/ft) is most nearly:

A) 2,386.2
B) <u>3,112.5</u>
C) 3,838.7
D) 4,565.0

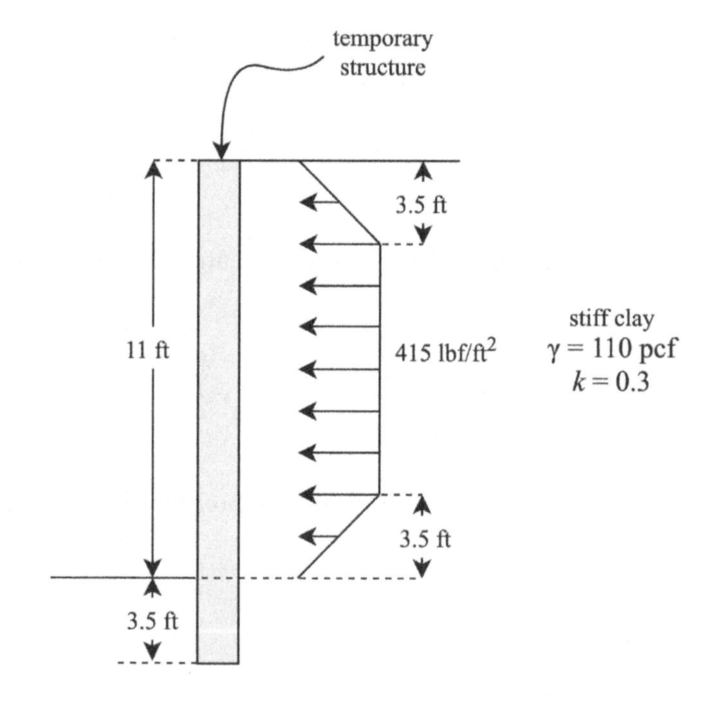

FIND: Total earth force per foot of wall (lbf/ft)

Step 1) Simply, calculate the resultant force:

$$R=2\left(\frac{1}{2}(3.5ft)\left(415\frac{lbf}{ft^2}\right)\right)+(4ft)\left(415\frac{lbf}{ft^2}\right)=3,112.5\frac{lbf}{ft}$$

Notes:

- The first part of the equation accounts for the two triangular forces, and the latter part for the middle force.
- The unit weight and k_a is given in this problem but is not used because the load is given. If the loads are not given, then use the unit weight and k_a to solve for the maximum load.

22) A crew is able to place 35 yd^3 of concrete per 8-hour work day. The project requires 2,098 ft^3 of concrete to be placed. Most nearly, what is the total cost ($) of the personnel involved in concreting if the crew consists of one foreman at $65/hr and three laborers at $32/hr?

A) 2,737
B) 2,860
C) 4,364
D) 8,578

FIND: Cost of personnel ($)

Step 1) Convert cubic feet to cubic yards:

$$2,098\ cf\left(\frac{yd}{3\,ft}\right)^3 = 77.70\ cy$$

Step 2) Find the cost of the whole crew per hour:

$$\left(1\ foreman\ x\ \frac{\frac{\$\,65}{hr}}{foreman}\right)+\left(3\ laborers\ x\ \frac{\frac{\$\,32}{hr}}{laborer}\right)=\frac{\$\,161}{hr}$$

Step 3) Find the number of hours required for the crew to work:

$$77.70\ cy\ x\frac{work\ day}{35\ cy}=2.22\ work\ days$$

Step 4) Calculate how many hours it will take to complete the project:

$$2.22\ work\ days\ x\frac{8\ hrs}{work\ day}=17.76\ hours$$

Step 5) Determine the total cost for the personnel:

$$17.76\ hrs\ x\ \frac{\$\,161}{hr}=\$\,2,859.36$$

Notes:

- Similar to rounding up on quantities, it's also prudent to round up on personnel costs because rounding down wouldn't cover the full payment.

23) A new office foundation is being constructed that is 95 ft x 100 ft. Soil must be replaced to a depth of 7.6 ft directly below the foundation. The soil will be borrowed from a nearby pit. Formulas are given for various excavation soil factors. The loose volume of soil (yd³) that is needed to be transported from the borrow pit to the new foundation is most nearly:

Quantity	Symbol	Units	Formulas		
bank volume	BCY	yd³	$\frac{LCY}{swell\,factor}$	$\frac{CCY}{(1-shrinkage)}$	
loose volume	LCY	yd³	$Shrinkage\ factor\ x\ BCY$	$\frac{BCY}{load\,factor}$	$\frac{\gamma_{bank}}{\gamma_{loose}} x\ BCY$

A) 2,674
B) 2,971
C) 3,118
D) 3,120

γ_{borrow} = 95 pcf
Swell = 5%
Shrinkage = 10%

FIND: Loose soil volume, LCY (yd³)

Step 1) Calculate the volume of compacted soil, V_C, for under the foundation. Note that we already know the space the compacted soil will fill and therefore don't need any conversions to determine V_C.

$$V_C = \frac{(100\,ft)(95\,ft)(7.6\,ft)}{\frac{27\,ft^3}{yd^3}} = 2,674\ yd^3$$

Step 2) Shrinkage is 10%, so the shrinkage factor is 0.10. Calculate the bank volume, V_B, required from the borrow site:

$$BCY = \frac{CCY}{1-DF} = \frac{2,674\ yd^3}{1-0.10} = 2,971\ yd^3$$

Step 3) Swell is 5%, so the swell factor, SF, is 1.05. Calculate the loose volume of the soil, LCY:

$$\begin{aligned} LCY &= (SF)(BCY) \\ &= (1.05)(2,971\ yd^3) \\ &= 3,119.55\ yd^3 \end{aligned}$$

Notes:

- CERM Chapter 80
- Earth work is always calculated in cubic yards, or simply just "yards".

24) An 11 ft cut in sandy soil is being supported by a reinforced concrete wall. The backfill is level and a surcharge pressure of 500 lbf/ft^2 extends behind the wall. Information about the soil is given and a diagram is shown below. If the surcharge resultant acts halfway up from the base and the active soil resultant acts one-third up from the base, the total overturning moment (lbf) per foot of wall, taken about the toe, is most nearly:

A) 5,765
B) <u>56, 211</u>
C) 72,787
D) 145,573

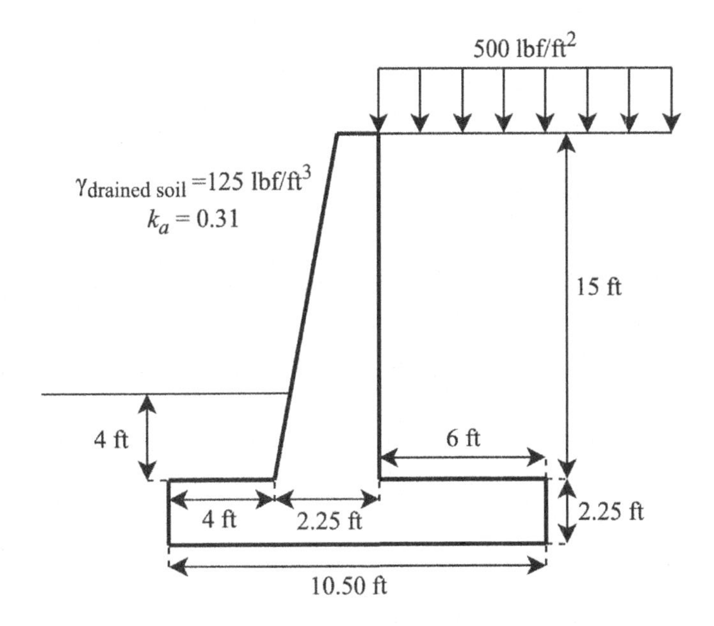

FIND: Overturning moment, M_{OT} (lbf)

Step 1) Search *surcharge* and navigate to *Chapter 3.1.4 Load Distribution from Surcharge* in the Handbook. Calculate the surcharge reaction, R_q, due to a uniform surcharge at a given depth:

$$R_q = k_a qHw = (0.31)\left(500\,\frac{lbf}{ft^2}\right)(17.25\,ft)(1\,ft) = 2,673.75\frac{lbf}{ft}$$

Step 2) Calculate the active soil resultant using the equation given for active force in the diagram entitled Failures Surfaces, Pressure Distribution and Forces: (a) Active case, (b) Passive case in Section 3.1.2 Rankine Earth Coefficients of the Handbook:

$$R_a = \frac{1}{2}k_a\gamma H^2 = \left(\frac{1}{2}\right)(0.31)\left(125\frac{lbf}{ft^3}\right)(17.25\,ft)^2 = 5,765.27\frac{lbf}{ft}$$

Step 3) Since the soil above the heel is horizontal, R_a is horizontal and there is no vertical component to R_a. As is given in the problem, the surcharge resultant acts halfway up from the base (H/2) and the active soil resultant acts one-third up from the base (H/3). Calculate the overturning moment, M_{OT}:

$$M_{OT} = R_q\left(\frac{H}{2}\right) + R_{a,h}\left(\frac{y_{a,h}}{3}\right)$$

$$= \left(2,673.75\frac{lbf}{ft}\right)\left(\frac{17.25\,ft}{2}\right) + \left(5,765.27\frac{lbf}{ft}\right)\left(\frac{17.25\,ft}{3}\right) = 56,211.40\ lbf\ (per\ foot\ of\ wall)$$

Notes: CERM Chapter 37

25) Which one of the following is most accurate regarding steel column design?

A) <u>As a column becomes longer, the load that causes the buckling becomes smaller.</u>
B) As a column becomes wider, the load that causes the buckling becomes smaller.
C) As a column becomes longer, the load that causes the buckling becomes larger.
D) As a column becomes shorter, the load that causes the buckling becomes smaller.

This can be determined by evaluating the equation in the Handbook's *Chapter 1.6.8 Columns.*

Notes: CERM Chapter 61

26) Which of the following equations describes the OSHA Incidence Rate?

A) <u>(Number of injuries, illnesses, and fatalities) x (200,000) / (Total hours worked by all employees during the period in question)</u>
B) (Total number of recordable incidents resulting in days away, restricted, or transferred) x (200,000) / (Total hours worked by all employees)
C) (Total number of lost time incidents) x (200,000) / (Total number of employees worked each year)
D) (Total number of recordable incidents) x (200,000) / (Total number of hours worked by injured employees)

Search *incidence rate* and navigate to *Chapter 2.6.1.1 Safety Incidence Rate.*

Notes:

- Option B is the *DART Rate*l Option C is the *Lost Time Injury Frequency Rate;* Option D is the *Total Recordable Incident Rate.*
- The 200,000 number is a benchmark established by OSHA to which a company can compare their own work hours. It represents 100 employees working 40-hour weeks for 50 weeks in a year.

27) Water is carried through a 311 ft long, 8 in. (internal diameter) schedule 40 welded and seamless pipe (Ɛ=0.0002 ft) that contains one fully open flanged steel angle valve, two flanged steel regular 90° elbows, and one flanged steel gate valve. The equivalent length of all combined fittings is 117.20 ft. The discharge is located 20 ft higher than the intake. Taking into consideration minor losses, the total head loss (ft) through the pipe is most nearly:

A) 361.70
B) 371.90
C) <u>381.70</u>
D) 392.00

$Re = 1.92 \times 10^5$
$v = 45$ ft/s

FIND: Total head loss, Δh (ft)

Step 1) Search *head loss* in the Handbook and navigate to the Darcy-Weisbach Equation in *Chapter 6.2.3.1 Head Loss Due to Flow*:

$$h_f = f\frac{Lv^2}{D2g}$$

Step 2) Search *moody diagram* and navigate to the Moody, Darcy, or Stanton Friction Factor Diagram, where you'll find that the specific roughness is 0.0002 ft for a welded and seamless steel pipe. Calculate the relative roughness

$$\frac{\epsilon}{D} = \frac{0.0002\,ft}{8\,in\left(\frac{ft}{12\,in}\right)} = 0.0003$$

Step 3) Using Re=1.92 x 10^5 and $\epsilon/D = 0.0003$, determine from the Moody diagram that $f = 0.0180$.

Step 4) Calculate the total equivalent pipe length: $L_{total} = L + L_{fittings} = 311\,ft + 117.2\,ft = 428.20\,ft$

Step 5) Compute the head loss due to friction:

$$h_f = f\frac{Lv^2}{D2g} = (0.0180)\frac{(428.20\,ft)\left(45\,\frac{ft}{s}\right)^2}{(0.67\,ft)(2)\left(32.2\,\frac{ft}{s^2}\right)} = 361.73\,ft$$

Step 6) Calculate the total head loss:

$$\Delta h = h_f + \Delta z = 361.73\,ft + 20\,ft = 381.73\,ft$$

Notes:

- CERM Chapter 17
- Watch out for *nominal* versus *actual* internal pipe diameters. A problem should specify "nominal", in which case converting to *actual* may be necessary. If a problem does not specify, then differentiation is likely not required.

28) Which of the following is the most appropriate friction factor to use in calculating earth pressure for a foundation wall interfacing with stiff clay?

A) <u>0.32</u>
B) 0.70
C) 0.22
D) 1.10

Navigate to the table in the Handbook entitled Wall Friction and Adhesion for Dissimilar Materials.

29) Soil was extracted from a depth of 7 meters. The grain size distribution characteristics, and liquid and plastic limits of the soil are as shown below. What is the classification of the soil according to the USCS?

A) SP
B) SW
C) CH
D) SC

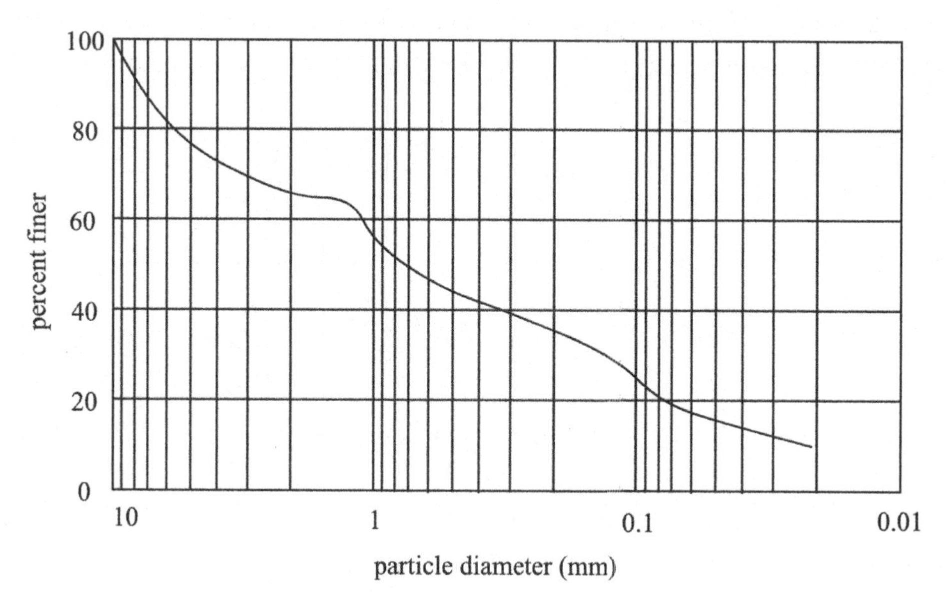

LL = 33.1
PL = 10.9

FIND: Soil classification

Step 1) Search *USCS* in the Handbook and navigate to *Chapter 3.7.2 Unified Soil Classification System.*

Step 2) Draw lines on the graph to find the percent finer than the #4 (4.5 mm) sieve and the #200 (0.075 mm) because these allow us to determine the *major division* of the soil in the first column of the Plasticity Chart.

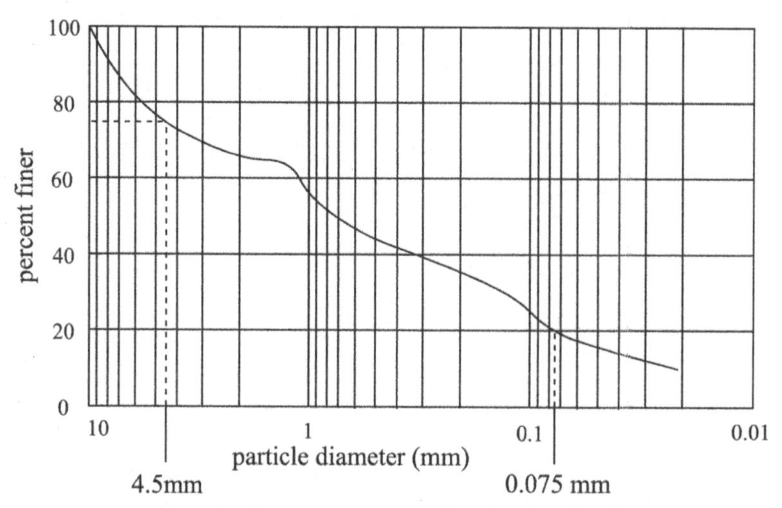

Step 3) First major division: The percentage finer than 0.075 mm (passing the #200 sieve) is 20%, which means that over 50% is coarser than the #200 sieve and therefore *coarse-grained* according to the Soil Classification Chart.

Step 4) Second major division: The percentage finer than the 4.5 mm (passing the #4 sieve) is 75% and therefore can be preliminarily classified on the Soil Classification Chart as sandy soil.

Step 5) Next we use the Plasticity Chart:

$$PI = LL - PL = 33.1 - 10.9 = 22.2 \text{ (Plasticity Index Equation)}$$

Step 6) These put us above the A-line on the plasticity chart. Per the fifth column in the table under supplementary requirements, the fines classify as CL or CH. Therefore the soil is clayey sand (SC).

Notes:

- CERM Chapter 35
- Be aware of the *and/or* options (i.e. "PI less than 4 or below A-line vs. PI over 7 and above A-line") in the Soil Classification Chart. "And" means you satisfy BOTH conditions!
- The coarse fraction is that fraction of the soil particles having grain sizes larger than a No. 200 sieve, or the fraction that is retained on the #200 sieve.
- Pay attention to whether you're dealing with *passing* or *retained.*
- The soil classification tables can also be used to eliminate the wrong answers instead of finding the right answer, if this method is faster for you.

30) The layout of a wall is shown. Allow for 4% brick waste. Mortar joints are required to be 0.5 in. thick. Two rows of brick positioned as stretchers are required. Using nominal sizing, most nearly, how many standard size non-modular bricks are needed?

A) 369
B) 370
C) 739
D) <u>769</u>

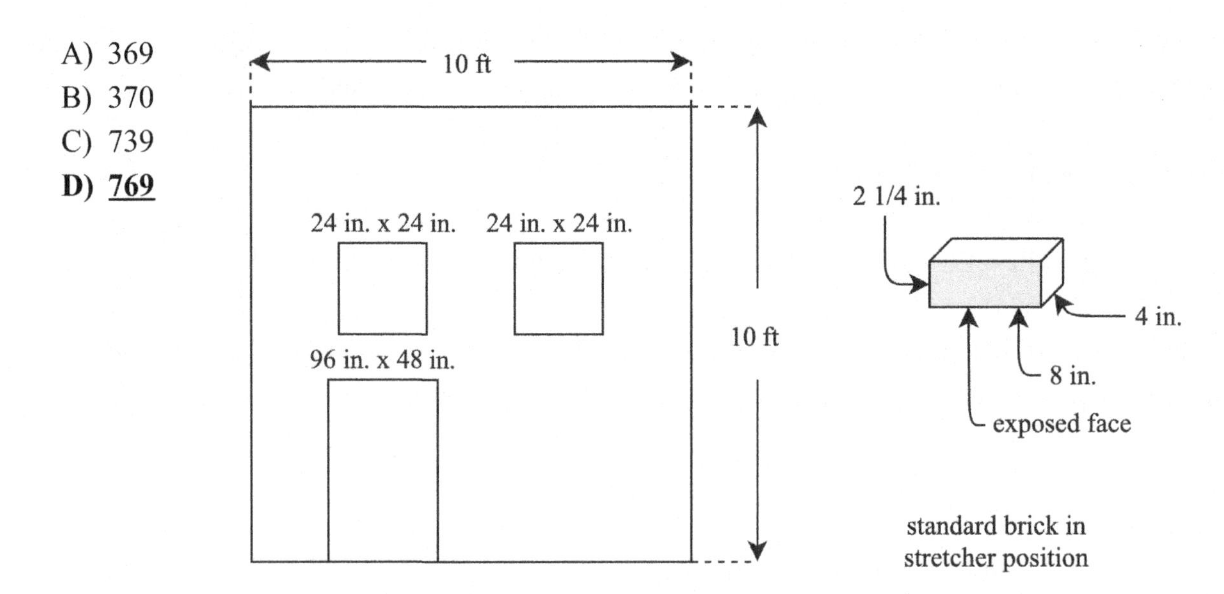

FIND: Number of bricks

Step 1) Convert the door and window sizes to feet, and calculate the net surface area of the wall:

$$(10ft \times 10ft) - 2(2ft \times 2ft) - (8ft \times 4ft) = 60ft^2$$

Step 2) Each mortar joint is 0.5 in., so each single brick carries 0.25 in. of mortar on each side. Therefore, we can add 0.5 in. to the brick's length and height respectively, and determine the area of the brick face with the mortar:

$$(8\ in. + 0.5\ in.) \times (2.25\ in. + 0.5\ in.) = 23.375\ in.^2$$

Step 3) Convert this to feet:

$$\left(\frac{23.375\ in^2}{brick}\right)\left(\frac{ft}{12\ in.}\right)^2 = \frac{0.162\ ft^2}{brick}$$

Step 4) Divide the net wall area by the surface area of a brick:

$$\frac{60\ ft^2}{\frac{0.162\ ft^2}{brick}} = 369.63\ bricks$$

Step 5) Multiply the number of rows by bricks required:

$$369.63\ bricks\ x\ 2\ rows\ =\ 739.25\ bricks$$

Step 6) Account for 4% waste:

$$739.25\ bricks\ x\ 1.04\ =\ 768.82\ bricks\ (769\ bricks)$$

31) What is the slope of the initial linear portion of the curve on the stress-strain diagram for steel?

A) Toughness
B) Modulus of Elasticity
C) Yield stress
D) Shear modulus

Search *stress-strain* in the Handbook and navigate to *Chapter 1.6.1 Uniaxial Stress-Strain.*

32) A 2-hr storm over a 95 km^2 watershed area produces a total runoff volume of 4.5x10^6 m^3 with a peak discharge of 325 m^3/s. If a 2-hr storm producing 8 cm of runoff is to be used to design a culvert in the same area, the design discharge (m^3/s) is most nearly:

A) 55
B) 69
C) 324
D) 553

FIND: Design discharge, Q (m^3/s)

Option 1

Step 1) Calculate the average depth of the excess precipitation, $P_{ave,\ excess}$:

$$P_{ave,\ excess} = \frac{V}{A_d} = \frac{\left(4.5x10^6\ m^3\right)}{\left(95\ km^2\right)\left(1{,}000\frac{m}{km}\right)^2} = 0.047\ m\ (4.7\ cm)$$

Step 2) Calculate the unit hydrograph discharge by dividing the peak discharge by the average precipitation:

$$Q_{unit\ hydrograph} = \frac{Q_p}{P_{ave,\ excess}} = \frac{325\ \frac{m^3}{s}}{4.7\ cm} = 69.15\frac{m^3}{s \cdot cm}$$

Step 3) Calculate the design discharge by multiplying the unit hydrograph discharge by the design 8 cm runoff:

$$Q_p = Q_{unit\ hydrograph}\ (8\ cm) = \left(69.15\frac{m^3}{s \cdot cm}\right)(8\ cm) = 553\frac{m^3}{s}$$

Option 2

Step 1) Same as above.

Step 2) Find the design discharge by directly comparing the unit hydrograph discharge to the desired design discharge:

$$\frac{4.7\ cm}{325\ \frac{m^3}{s}} = \frac{8\ cm}{x\frac{m^3}{s}}$$

$$x = 553\frac{m^3}{s}$$

Notes:

- CERM Chapter 20
- Units of the unit hydrograph are in/in (or cm/cm). Therefore, the units for $Q_{\text{unit hydrograph}}$ are Q *per* inch (i.e. ft^3/s-inch), which is typically left out because it is implied for the unit hydrograph.
- Excess precipitation is the volume of rainfall available for direct surface runoff. It is equal to the total amount of rainfall minus all abstractions including infiltration, interception, and depression storage. Because runoff is the excess precipitation after abstractions, "excess" and "total" are generally interchangeable.

33) The saturated unit weight (kN/m^3) of a collected soil sample with the given characteristics is most nearly:

A) 17.66
B) 17.76
C) 19.58
D) 19.68

SG = 2.67
$e_{max} = 0.92$
$e_{min} = 0.34$
$D_r = 44\%$

FIND: Saturated unit weight (kN/m^3)

Step 1) As you're given the relative density, D_r, search *relative density* in the Handbook and navigate to *Chapter 5.5.1 Relative Soil Density.*

Step 2) Using the equation for relative density, determine the void ratio, *e*:

$$D_r = \frac{(e_{max} - e)}{(e_{max} - e_{min})}$$

$$\Rightarrow e = -D_r(e_{max} - e_{min}) + e_{max}$$

$$= -0.44(0.92 - 0.34) + 0.92 = 0.66$$

Step 3) Search *saturated unit weight* in the Handbook and navigate to *Chapter 3.8.3 Weight-Volume Relationships*. Find the appropriate equation in the Weight-Volume Relationships table and solve:

$$\gamma_{sat} = \frac{(SG + e)\gamma_w}{1 + e}$$

$$= \frac{(2.67 + 0.66)\left(9.81 \frac{kN}{m^3}\right)}{1 + 0.66} = 19.68 \frac{kN}{m^3}$$

Notes:

- CERM Chapter 35

34) It takes 7 minutes and 45 seconds for a 30 mm thick two-way drained clay layer to consolidate 30% in the lab. Most nearly, how many days are required for a 9 m thick two-way drained layer of the same clay to reach 30% consolidation in the field?

A) 8
B) **<u>484</u>**
C) 501
D) 1,938 **FIND: Time to reach 30% consolidation in field (days)**

<u>Option 1</u>
Step 1) Search *consolidation* in the Handbook and navigate to *Chapter 3.2.3 Time Rate of Settlement.* Find the appropriate equation and rearrange to solve for the field consolidation:

$$t = \frac{T_v (H_d)^2}{c_v}$$

$$T_{30\%} = \frac{c_v t_{lab}}{H^2_{lab}} = \frac{c_v t_{field}}{H^2_{field}}$$

$$t_{field} = H^2_{field} \frac{t_{lab}}{H^2_{lab}} = \left(\frac{9\ m}{2}\right)^2 \frac{7.75\ min\left(\frac{hr}{60\ min}\right)\left(\frac{day}{24\ hr}\right)}{\left(\left(\frac{30\ mm}{2}\right)\left(\frac{1\ m}{1000\ mm}\right)\right)^2} = 484.38\ days$$

<u>Option 2</u>
Step 1) Begin with the same equation as Option 1, and rearrange to solve for c_v. Plug in lab results and T_v of 0.071 (corresponds with U_z of 30%) from the Average Degree of Consolidation versus Time Factor table:

$$t = \frac{T_v (H_d)^2}{c_v} \Rightarrow c_v = \frac{T_v (H_d)^2}{t} = \frac{(0.071)\left(30\ mm \cdot \frac{m}{1000\ mm}\right)^2}{7.75\ min \cdot \frac{hr}{24\ min} \cdot \frac{day}{24\ hr}} = 0.0119 \frac{m^2}{day}$$

Step 2) Plug c_v back into the original equation using the field conditions:

$$t = \frac{T_v (H_d)^2}{c_v} = \frac{0.071\ (9\ m)^2}{0.0119 \frac{m^2}{day}} = 484.37\ days$$

Notes: CERM Chapter 40

35) A 48-km² watershed area is shown. Rain gauge stations and their recorded rainfall values are marked. The isohyets have also been drawn. The approximate mean precipitation (mm) of the catchment area is most nearly:

A) 2.79
B) 3.07
C) 5.30
D) <u>6.88</u>

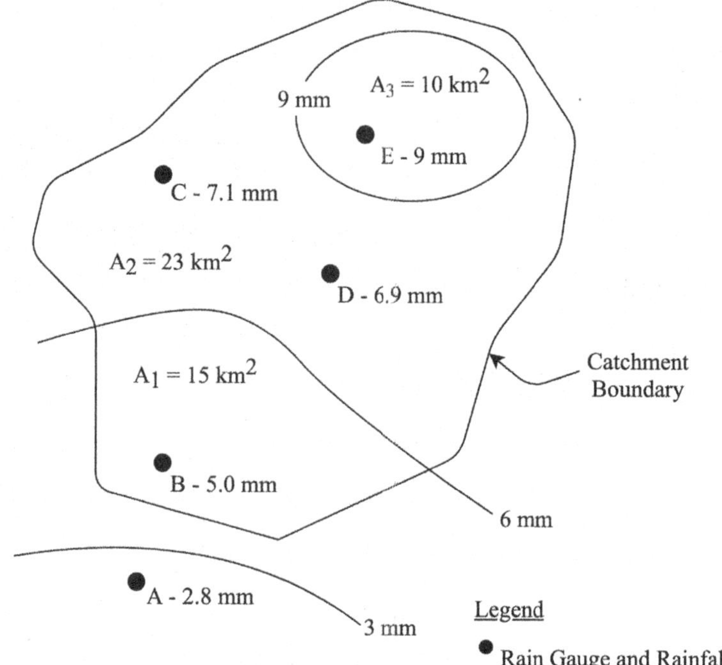

FIND: Mean precipitation (mm)

Step 1) Search *isohyetal method* in the Handbook and navigate to *Chapter 6.5.6.3 Isohyetal Method.* Calculate the mean precipitation where A_i is the area between the isohyets and P_i is the isohyet rainfall value:

$$\overline{P} = \frac{1}{A}\sum_{i=1}^{n} A_i P_i$$

$$\overline{P} = \frac{\left(15km^2\left(\frac{3mm+6mm}{2}\right) + 23km^2\left(\frac{6mm+9mm}{2}\right) + 10km^2 \cdot 9mm\right)}{48\ km^2}$$

$$= 6.88\ mm$$

Notes:

- Regarding the isohyetal method, the station data are used to draw isohyets, but they are not used in the calculation of average rainfall. As is shown in this problem, the rain gauge data is given but the isohyets (showing constant precipitation) are used to calculate average precipitation per area.
- We recommend also briefly studying other methods such as the Thiessen Method.

36) Which of the following is the most accurate description of moderately weathered rock?

A) Rock shows slight discoloration.
B) More than half of the rock is decomposed.
C) Less than half of the rock is decomposed.
D) Original minerals of rock have almost entirely decomposed.

Search *rock weathering* and navigate to the table in Chapter 3 of the Handbook entitled Terms to Describe Rock Weathering and Alteration.

37) The section modulus, S, of a rectangular b x h section is calculated as: $S_{rectangular} = bh^2/6$. The maximum bending stress at midspan (kN/m^2) for the simply supported beam shown below is most nearly:

A) 48
B) 64
C) 128
D) <u>192</u>

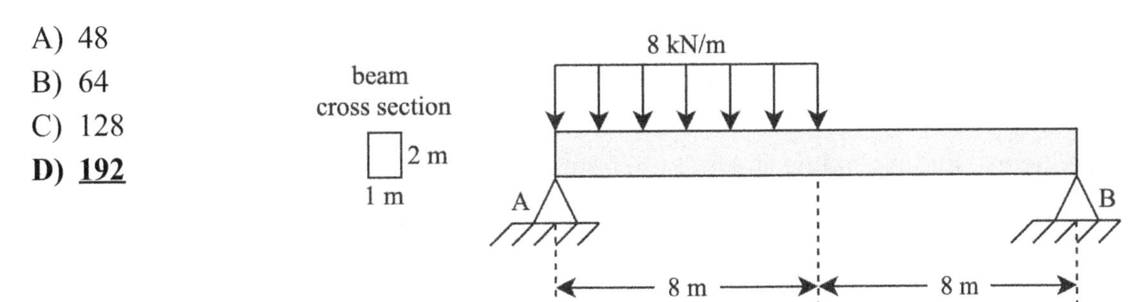

FIND: Maximum bending stress at midspan, $\sigma_{b,\ max,\ midspan}$ (kN)

Step 1) Search *bending stress* in the Handbook and navigate to *Chapter 1.6.7.2 Stresses in Beams*. Find the equation for the maximum normal bending stress and insert the given equation for the section modulus of a rectangular section:

$$\sigma_{b,\ max} = \frac{M}{S} = \frac{M}{\left(\frac{bh^2}{6}\right)}$$

Step 2) Simplify the beam into its components:

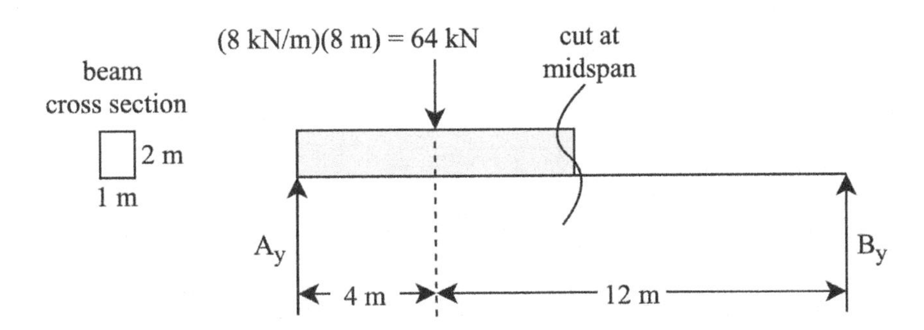

No horizontal forces act on the beam. The two reaction forces are the vertical forces at supports A (R_A) and B (R_B).

Step 3) Solve for R_A by setting the moment around support B to zero:

$$M_B = \sum Fd = 0$$

$$= R_A(16\ m) - (64\ kN)(12\ m) = 0$$

$$R_A = \frac{(64\ kN)(12\ m)}{16\ m} = 48\ kN$$

Step 4) Calculate the moment at midspan:

$$M_{mid} = \sum Fd = 0$$

$$= (48\ kN)(8\ m) - (64\ kN)(4\ m) = 128\ kN \cdot m$$

Step 5) Calculate the maximum bending stress at midpoint, $\sigma_{b,max,\ midpoint}$:

$$\sigma_{b,\ max,\ midspan} = \frac{M_{mid}}{S} = \frac{M_{mid}}{\left(\frac{bh^2}{6}\right)} = \frac{128\ kN \cdot m}{\frac{(1m)(2m)^2}{6}} = 192 \frac{kN}{m^2}$$

Notes: CERM Chapter 44

38) A 2:1 (horizontal:vertical) sloped cut is made in submerged clay. The cut is 38 ft deep and the clay extends 19 ft below the toe of the cut before it hits a rock layer. The cohesive factor of safety of the cut is 1.5. The saturated density of the clay (lbf/ft^3) is most nearly:

A) 54
B) 132
C) 178
D) 194

$c = 1{,}200\ lbf/ft^2$
$F_{cohesive} = 1.5$

FIND: Saturated density of the clay, γ_{sat} (lbf/ft^3)

Step 1) Search *slope stability* in the Handbook and navigate to *Chapter 3.6.1 Stability Charts*. Find the Factor of Safety equation in the Stability Number chart.

$$F = N_o \frac{c}{\gamma H} \quad \Rightarrow \gamma = \frac{N_o c}{FH}$$

Step 2) Calculate the slope angle, β:

$$\beta = arctan(1/2) = 26.6°$$

Step 3) Use the equation within the Stability Number chart to calculate the depth factor, d:

$$d = \frac{D}{H} = \frac{19\,ft}{38\,ft} = 0.50$$

Step 4) Deduce from the Stability Number chart that $N_o \cong 6.25$.

Step 5) Calculate γ (γ_{eff}):

$$\gamma_{eff} = \frac{N_o c}{FH} = \frac{(6.25)\left(1,200\frac{lbf}{ft^2}\right)}{(1.5)(38\,ft)} = 131.58\frac{lbf}{ft^3}$$

Step 6) Calculate γ_{sat} :

$$\gamma_{sat} = \gamma_{eff} + \gamma_{water} = 131.58\frac{lbf}{ft^3} + 62.40\frac{lbf}{ft^3} = 193.98\frac{lbf}{ft^3}$$

Notes: CERM Chapter 40

39) A road is required to have a 3% crown to properly drain stormwater runoff to the gutters. The gutter detail is shown below. If the centerline of the road is 4783.14 ft and the road is 34 ft wide (TBC to TBC), the elevation of the flow line (ft) is most nearly:

A) 4781.38
B) <u>4782.61</u>
C) 4782.63
D) 4782.69

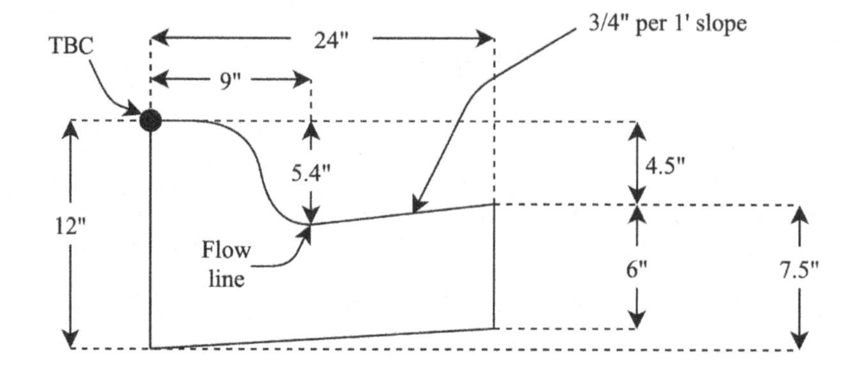

FIND: Elevation of flow line (ft)

Step 1) Calculate the distance from the centerline to the edge of the curb:

$$Edge\ of\ Curb\ to\ Edge\ of\ Curb = 34\,ft - 2(2\,ft) = 30\,ft$$

$$Centerline\ to\ Edge\ of\ Curb = \frac{30\,ft}{2} = 15\,ft$$

$$Elevation_{edge\ curb} = 4783.14\,ft - 15\,ft\,(0.03) = 4782.69$$

Step 2) Calculate the elevation at the flow line.

The distance between the edge of the curb and flow line is 24" - 9" = 15". If the slope is ¾" (0.0625') per 1', then the slope is 6.25%. Therefore, the elevation at the flow line is:

$$Elevation_{flow\ line} = 4782.69 - 15\,in\left(\frac{ft}{12\,in}\right)(.0625) = 4782.61$$

40) The structure shown below carries a dead load of 45 psf (including self weight) and a live load of 55 psf. The unfactored axial stress (ksi) in the center column is most nearly:

A) 0.25
B) 0.28
C) <u>1.50</u>
D) 1.55

$A_{column} = 8\ in^2$

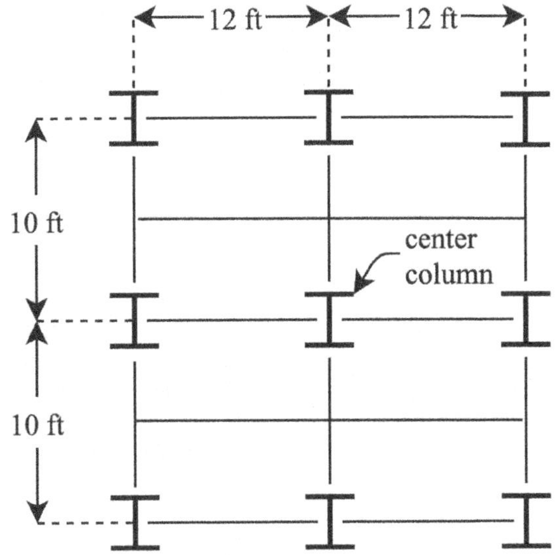

12 ft 12 ft
10 ft
10 ft
center column

FIND: Axial stress, σ (ksi)

Step 1) Search *axial loading* in the Handbook and navigate to *Chapter 1.6.2.6 Uniaxial Loading and Deformation* where you'll find the equation for stress on the cross section:

$$\sigma = \frac{P}{A}$$

Step 2) Calculate the tributary area of the center column:

$$A_{trib} = \frac{(12ft + 12ft)(10ft + 10ft)}{4} = 120\,ft^2$$

Step 3) Calculate the total load on the column:

$$P = A_{trib}(DL + LL)$$

$$= 120\,ft^2\left(45\frac{lbf}{ft^2} + 55\frac{lbf}{ft^2}\right)$$

$$= 12,000\ lbf = 12\ kips$$

Step 4) Calculate the axial stress:

$$\sigma = \frac{F}{A} = \frac{12\ kips}{8\ in^2} = 1.5\ ksi$$

Notes: CERM Chapter 44

<u>- END OF SOLUTIONS VERSION 2 -</u>

1) The station of the PT in the horizontal curve shown below is most nearly:

A) 12+88.53
B) 25+43.82
C) 33+46.53
D) 37+97.82

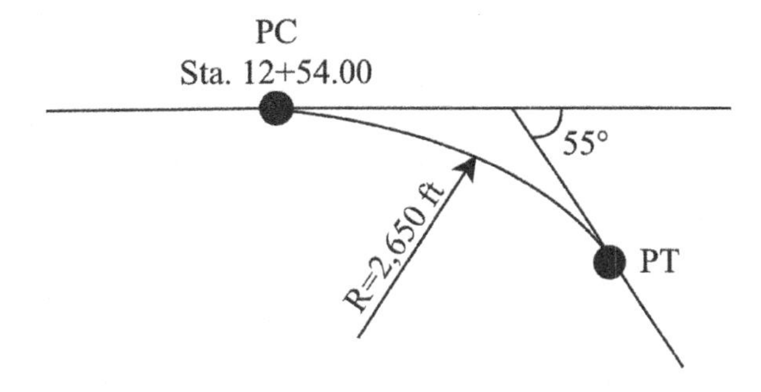

2) Before development, a 0.05 square mile watershed had two different sections with the following sizes and runoff coefficients: ⅓ of the watershed area with C=0.20 and ⅔ of the watershed area with C=0.35. After development, the watershed has a uniform runoff coefficient of 0.85. A detention pond must be sized to discharge at the pre-development rate. Based on a 25-yr storm with an intensity of 0.25 in/hr and ignoring the stage-discharge relationship, the design discharge (cfs) for the pond's outlet is most nearly:

A) 2.4
B) 4.4
C) 4.8
D) 9.2

3) A sample of saturated clay from a consolidometer test has a total mass of 2.97 lbm and a dry mass of 2.55 lbm. The solid particles have a specific gravity of 2.5. The porosity of the sample is most nearly:

A) 0.16
B) 0.29
C) 0.58
D) 0.74

4) A water treatment facility is assessing the feasibility of a chlorination system replacement with a lifetime of 20 years. What is the benefit-cost ratio of the replacement and is it expected to deliver a positive net present value?

A) 0.58, no
B) 1.44, no
C) 1.39, yes
D) 1.73, yes

$i = 5\%$
n = 20 years

	Existing	**Replacement**
Initial Investment:	0	\$85,000
Salvage Value:	\$12,000	\$25,000
Annual Costs, 1-5 years:	\$10,000	\$3,000
Annual Costs, 6-20 years:	\$14,000	\$4,000

5) Which of the following is calculated by subtracting depreciation from initial cost?

A) Capitalized costs
B) Rate-of-return
C) Inflation
D) Book value

6) Which three Best Management Practices are most suitable for sediment control?

A) Silt fence, straw wattle, compacted earthen berm
B) Mulching, check dams, rock bags
C) Straw wattle, mulch blanket, geotextile filter fabric
D) Gradient terraces, hay bales, soil roughening

7) If no stratum is encountered above this depth, what is the minimum depth of exploration for embankment foundations?

A) Equal to twice the embankment height
B) Equal to a quarter of the embankment height
C) Between 1 and 2 times the wall height
D) 10 feet

8) A direct shear test is performed on a sample of sandy soil. Horizontal stress is slowly applied to the sample until it fails in shear at 40 kPa. Vertical stress of 95 kPa is applied to the sample. The soil sample's angle of internal friction (degrees) is most nearly:

A) 23
B) 24
C) 26
D) 35

9) A passenger car is traveling on a road made of portland cement concrete at a speed of 40 mph. The driver sees a stop sign and presses on the brakes in order to bring the car to a complete stop. The road is sloped downward at a 3.15% grade. The equation for stopping distance is provided. The coefficient of skidding friction, *f*, is 0.33. Most nearly, from how far away (ft) should the stop sign be visible?

$$s_{stopping} = vt_p + s_b = v_{1,\,mph}t_p + \frac{v^2_{1,\,mph} - v^2_{2,\,mph}}{30(f + G)}$$

A) 237
B) 294
C) 325
D) 8,979

10) A single-family residence is being constructed in an arid area with high seismic activity. Which construction type is most likely to be used?

A) Concrete bearing walls
B) Masonry bearing walls
C) Wood framing
D) Steel framing

11) A California Bearing Ratio, CBR, test is performed to determine the suitability of a soil for use as a subbase in pavement sections. If the field CBR is determined to be 70, what type of soil is this likely to be?

A) CL
B) SM
C) GP
D) GW

12) Most nearly, what is the velocity (fps) through the pipe for the system shown?

A) 12
B) 14
C) 37
D) 48

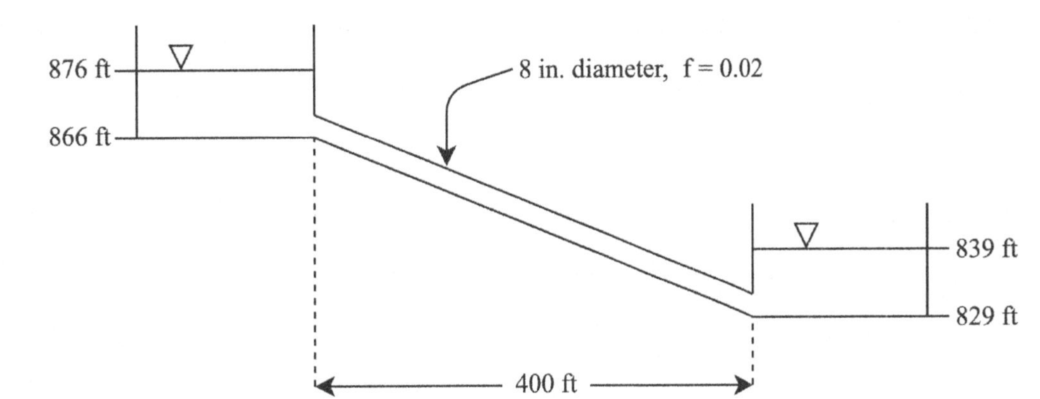

13) Which is the most correct equation to determine the ratio of total reinforcement area to the cross-sectional area of a concrete column?

A) Depth of equivalent rectangular stress block/area of shear reinforcement
B) Gross area of concrete section/total area of longitudinal reinforcement
C) Center-to-center spacing of torsional reinforcement/design flexural strength
D) Total area of longitudinal reinforcement/gross area of concrete section

14) What is the most typical way to repair a foundation that has been settled?

A) Underpinning
B) Retaining wall construction
C) Installation of french drains
D) Soil replacement

15) Web yielding or web crippling, two types of local buckling in steel beams, can most easily be reduced or eliminated by the use of:

A) Intermediate stiffeners and flange stiffeners
B) Bearing stiffeners and exterior enamel
C) Exterior enamel and flange stiffeners
D) Wooden exterior reinforcement and double thick beams

16) An activity-on-node diagram for a project is shown below with the activity durations in days. The float time (days) of Activity E is most nearly:

A) 12
B) 8
C) 10
D) 6

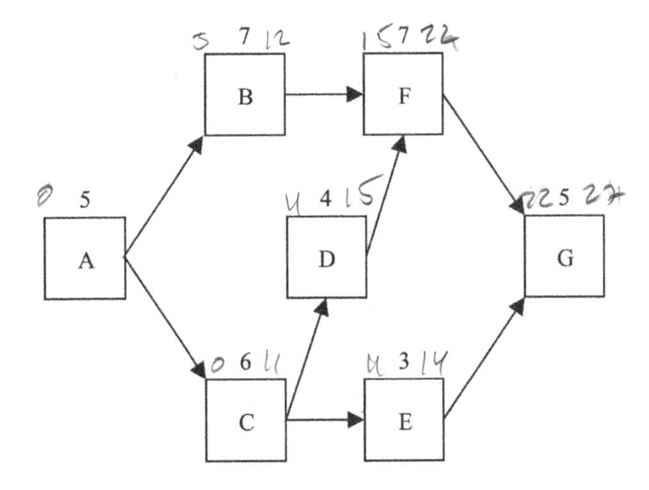

17) The following measurements were taken for four ring soil samples. The moisture content (%) of the sample is most nearly:

A) 7.95
B) 8.25
C) 15.15
D) 28.00

Total recovered mass of the four rings: 934 g
Average mass of empty ring sample container: 54 g
Average height of a single ring: 1.12 in
Dry unit weight of the soil: 58 lbs/ft^3
Average diameter of a ring sample: 3.51 in

18) An underpass is being designed in a city. Given the following diagram with the elevations shown, what is most nearly the elevation (ft) at point F?

A) 454.50
B) 498.50
C) 513.50
D) 511.00

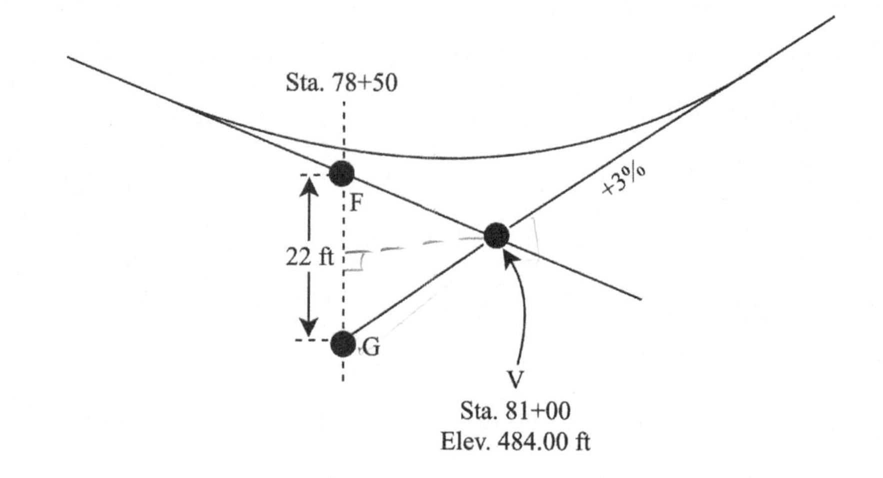

19) A hydraulic jump forms at the toe of a concrete rectangular spillway with uniform width. Before the jump, the water surface level is 8 ft above the apron with a velocity of 35 ft/s. The energy loss in the jump (ft) is most nearly:

A) 0.25
B) 3.25
C) 20.50
D) 21.25

20) Which of the following is usually not an applicable method for improving subgrade?

A) Using extruded geogrid when the subgrade California Bearing Ratio (CBR) = 5
B) Using woven geotextile when the subgrade CBR = 2
C) Using extruded geogrid when subgrade CBR = 1
D) Using woven geotextile when the subgrade is firm or stiff, and the total subbase thickness is 160 mm.

21) Which AASHTO specification is used to measure the relative proportion of plastic fines and dust to sand size particles in material passing the No. 4 sieve?

A) T176
B) D2419
C) C88
D) T210

22) An investor is considering the construction of a small water treatment facility with the expected lifetime of 35 yrs. Most nearly, what is the present value of the investment ($)?

Initial project cost = $900,000
Annual maintenance = $12,000
Salvage value = $90,000
Annual profit = $65,000
$i = 5\%$

A) 14,989.84
B) -15,850.40
C) 16,317.00
D) -196,190.40

23) Using the profile mass diagram shown, the total cut volume (yd^3) is most nearly:

A) 0.40×10^5
B) 2.0×10^5
C) 2.6×10^5
D) 200×10^5

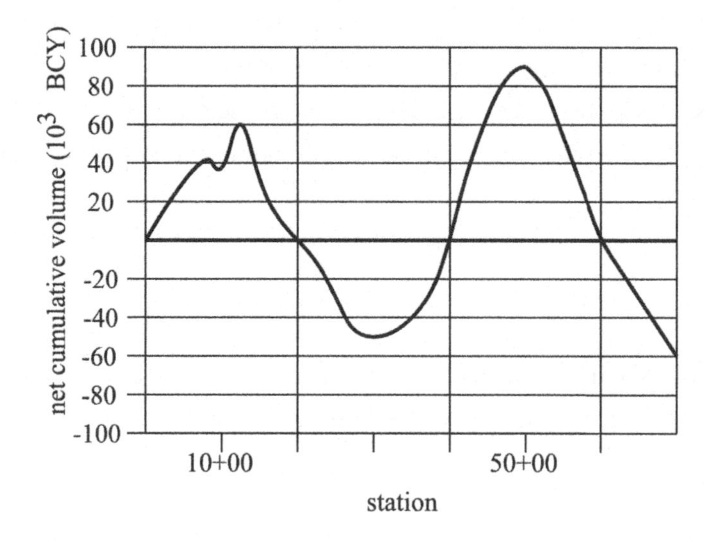

24) Which of the following statements is false regarding the relationship between lateral earth pressure and retaining walls?

A) Cohesive soils impose less pressure on retaining walls than cohesionless soils.
B) At-rest earth pressure imposes very nearly zero strain in the soil.
C) Passive earth pressure is the pressure in front of the wall that is a result of the wall moving towards the soil.
D) Active earth pressure is present behind a retaining wall that moves towards and compresses the remaining soil.

25) A fixed W12x96 column has a radius of gyration of 5.35 in. along the strong axis and 2.95 in. along the weak axis. The Euler buckling load of the column (kips) is most nearly:

A) 1,882
B) 1,992
C) 4,129
D) 4,264

$K = 1$
$E = 29x10^6$ psi
$A = 35.25$ in^2
$L = 18$ ft

26) During peak conditions, a 10 in water supply pipe line provides 0.67 cfs at the end of a 13,000 ft long supply line. What is the pressure head difference (ft) between the intake and discharge?

A) 125
B) 126
C) 132
D) 133

Intake elevation = 325 ft
Discharge elevation = 196 ft
C = 200

27) A developer is looking at building a 2 story building on a site in a mild climate that does not freeze. The boring log indicates that the upper 4 ft of the soil profile is sandy peat. The sandy peat is underlain by 38 ft of dense sand, followed by 21 ft of stiff clay. What is the most suitable type and depth of foundation for the building?

A) Shallow foundation at 1 ft
B) Shallow foundation at 5 ft
C) Pile foundation at 7 ft
D) Pile foundation at 32 ft

28) The foundation wall is to be 8 ft deep and 15 inches thick with a plan view shown. The volume of concrete (yd^3) required to be brought to the site for the foundation is most nearly:

A) 25
B) 64
C) 66
D) 67

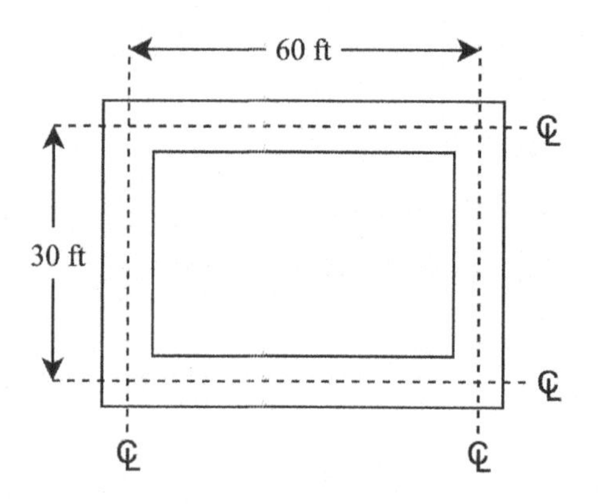

29) The watershed unit hydrograph of a 1-hr storm has a peak discharge of 225 cfs. If a 1-hr storm drops 1.75 in net precipitation, the peak discharge (cfs) is most nearly:

A) 11
B) 129
C) 130
D) 394

30) In analyzing a determinate statics problem, which configuration and orientation of forces describes one in which all the forces act at the same point?

A) Collinear force system
B) Singular point force system
C) Coplanar force system
D) Concurrent force system

31) A permeability test is conducted with a sample of soil that is 123 mm long and has a diameter of 64 mm. The flow is 1.45 mL in 484 seconds and the head is kept constant at 201 mm. The coefficient of permeability (m/yr) is most nearly:

A) 18.0
B) 18.5
C) 19.1
D) 19.2

32) The width of a reinforced concrete wall footing is 5 ft and the factored wall load per unit length is 25 kips. What is the factored load per unit length (kips/ft) at failure?

A) 5
B) 25
C) 15
D) 50

33) According to OSHA, what is the noise dose (%) of 90 dBa during a standard 8-hr work day? Does it require abatement?

A) 8.5%; No abatement required
B) 9.0%; Abatement required
C) 40%; No abatement required
D) 1600%; Abatement required

34) For determining slope stability for cohesive foundation soils, which of the following is the most likely reason for consolidating some samples to a higher than existing in situ stress?

A) To determine the increase of deflection needed to reduce the pore pressure to the required levels.
B) To determine the results of the field vane shear test.
C) To determine the shear strength in relation to depth.
D) To determine strength gain of clay due to consolidation under staged fill heights.

35) Two contiguous 4-ac watersheds shown below are served by an adjacent 500 ft long storm drain. Inlets are placed along the storm drain to collect runoff from the respective watershed. All flows are maximum. The storm drains flow full. The intensity of a storm is 5 in./hr. The flow time (min.) from inlet 1 to inlet 2 is most nearly:

A) 0.29
B) 4.78
C) 11.30
D) 41.80

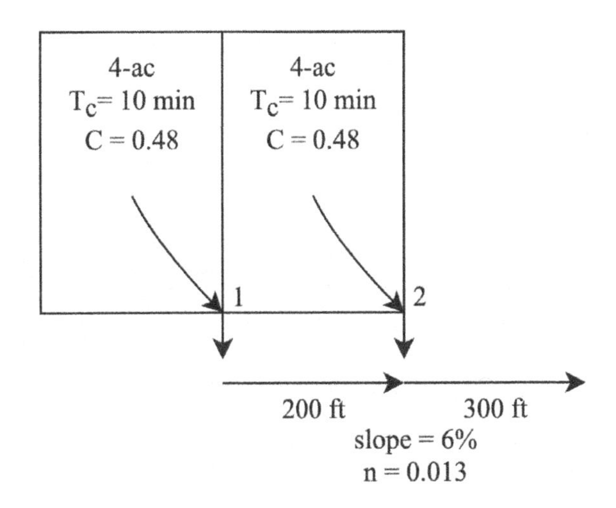

36) If the water table is at the bottom of Layer 1 as shown below, the total settlement (inches) of the normally consolidated clay layer for the soil profile shown below is most nearly:

A) 0.16
B) 0.23
C) 1.88
D) 2.76

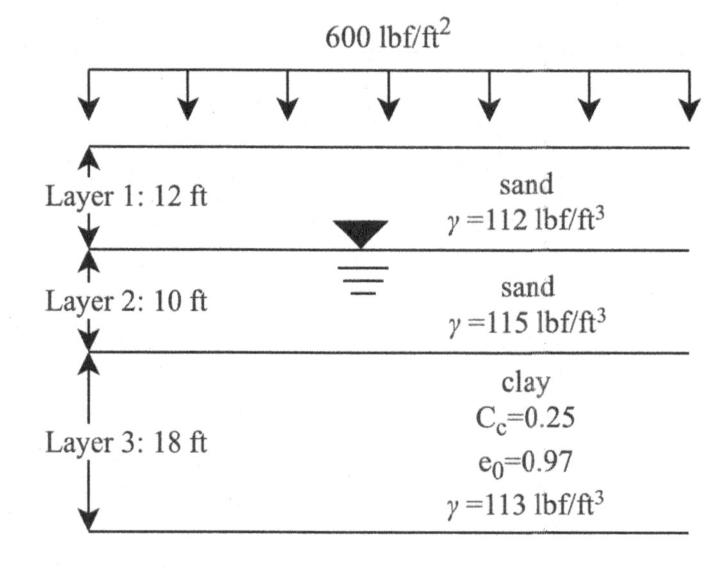

37) Which of the following gel times (min.) would you use for a low concentration silicate grout?

A) 200
B) 320
C) 440
D) 880

38) The 25 ft long Oak beam has a 2-in. x 8-in. cross section. The beam is subject to a uniform load of 0.7 kip/ft and a point load of 6.25 kips at 6 ft from Point A, as shown. Most nearly, where is the maximum moment measured from point A (ft)?

A) 4.36
B) 6.02
C) 10.36
D) 13.50

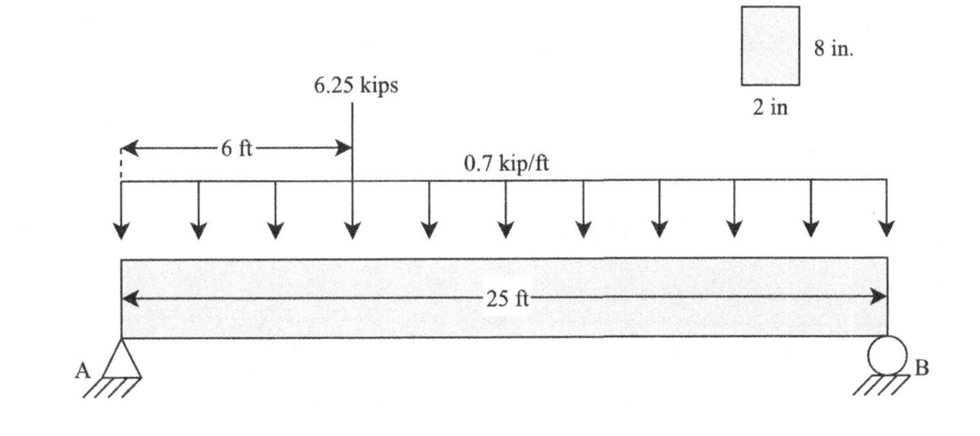

39) A public circular storm sewer is designed to carry a peak flow of 8 ft^3/s. The municipality requires that the depth at peak flow is a maximum of 75% of the pipe diameter. Based on topography, the required sewer slope is 2.5%. The required sewer diameter (in.) is most nearly:

$n = 0.012$

A) 14
B) 21
C) 18
D) 12

40) The equation for beam flanges in compression is given. A compact steel beam has the parameters shown. Most nearly, what is the required allowable stress yield (ksi) for the flanges to be in flexural compression?

A) 36
B) 50
C) 15
D) 18

$$\frac{b_f}{2t_f} \leq 0.38\sqrt{\frac{E}{F_y}}$$

Flange width = 24 inches
Flange thickness = ¾ inch
E = 29,000 ksi

FIND: Allowable stress yield, F_y (ksi)

- END OF EXAM VERSION 3 -

1) The station of the PT in the horizontal curve shown below is most nearly:

A) 12+88.53
B) 25+43.82
C) 33+46.53
D) <u>37+97.82</u>

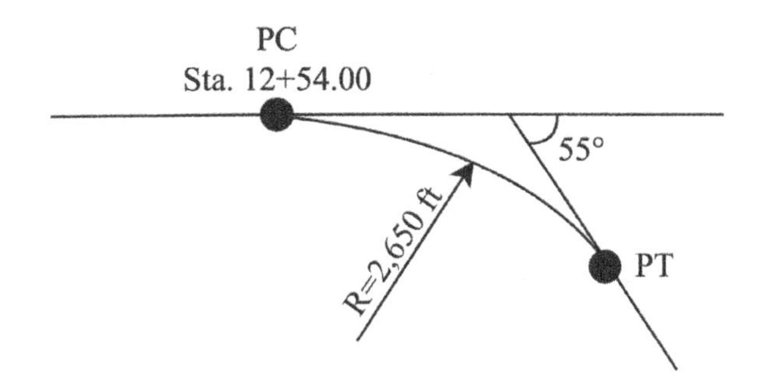

FIND: Station of the PT

Step 1) Search *horizontal curve* in the Handbook and navigate to *Chapter 5.2.1 Basic Curve Elements*. Determine from the Parts of a Circular Curve figure that Δ=55°. Calculate the length of the curve, L, using the given equation:

$$L = \frac{R \Delta \pi}{180}$$

$$= \frac{(2,650 ft)(55°)\pi}{180}$$

$$= 2,543.82 ft$$

Step 3) Calculate Station PT:

$$sta\ PT = sta\ PC + L$$

$$= 1,254.00 + 2,543.82$$

$$= 3,797.82$$

$$= 37 + 97.82$$

Notes:

- CERM Chapter 79
- Although the exam solution options should account for rounding differences, using π instead of an abbreviated value is always a safe bet.

2) Before development, a 0.05 square mile watershed had two different sections with the following sizes and runoff coefficients: ⅓ of the watershed area with C=0.20 and ⅔ of the watershed area with C=0.35. After development, the watershed has a uniform runoff coefficient of 0.85. A detention pond must be designed to discharge at the pre-development rate. Based on a 25-yr storm with an intensity of 0.25 in./hr and ignoring the stage-discharge relationship, the design discharge (cfs) for the pond's outlet is most nearly:

A) **2.4**
B) 4.4
C) 4.8
D) 9.2

FIND: Design discharge, Q (cfs)

Step 1) Search *rational method* in the Handbook and navigate to *Chapter 6.5.2.1 Rational Formula Method*:

$$Q = CIA$$

Step 2) Convert the area into acres and find the weighted runoff coefficient:

$$0.05\ sq\ mi \left(\frac{640\ ac}{sq\ mi}\right) = 32\ ac$$

Step 3) Calculate the weighted runoff coefficient:

$$C = \frac{\left(\frac{1}{3}\right)(32\ ac)(0.20) + \left(\frac{2}{3}\right)(32\ ac)(0.35)}{32\ ac} = 0.30$$

Step 4) Calculate the pre-development flow, which is the design discharge:

$$Q_{pre} = (0.30)\left(0.25\frac{in}{hr}\right)(32\ ac) = 2.40\ cfs$$

Notes:

- CERM Chapter 20
- Although this problem ignores it, understanding the stage-discharge curve is important.
- Know and understand the units involved in the various runoff method calculations, and how to find the weighted values.

3) A sample of saturated clay from a consolidometer test has a total mass of 2.97 lbm and a dry mass of 2.55 lbm. The solid particles have a specific gravity of 2.5. The porosity of the sample is most nearly:

A) 0.16
B) <u>0.29</u>
C) 0.58
D) 0.74

FIND: Porosity, *n*

Step 1) Search *moisture content* and navigate to the Volume and Weight Relationship table in *Chapter 3.8.3 Weight-Volume Relationships*. Find the moisture content equation:

$$w = \frac{W_w}{W_s} \cdot 100\% = \frac{2.97\ lb - 2.55\ lb}{2.55\ lb} \cdot 100\% = 16.47\%$$

Step 2) Because the clay is saturated, S=1. Determine the void ratio, *e*:

$$e = \frac{wG}{S} = \frac{0.1647 \cdot 2.5}{1} = 0.41$$

Step 3) Determine the porosity, *n*:

$$n = \frac{e}{1+e} = \frac{0.41}{1+0.41} = 0.29$$

Notes:

- CERM Chapter 35
- As you may have realized by now, this Volume and Weight Relationship table is very important so make sure you know how to find it quickly!

4) A water treatment facility is assessing the feasibility of a chlorination system replacement with a lifetime of 20 years. What is the benefit-cost ratio of the replacement and is it expected to deliver a positive net present value?

A) 0.58, no
B) 1.44, no
C) <u>1.39, yes</u>
D) 1.73, yes

$i = 5\%$
n = 20 years

	Existing	**Replacement**
Initial Investment:	0	$85,000
Salvage Value:	$12,000	$25,000
Annual Costs, 1-5 years:	$10,000	$3,000
Annual Costs, 6-20 years:	$14,000	$4,000

FIND: Benefit-cost ratio

Step 1) Search *benefit-cost ratio* in the Handbook and navigate to *Chapter 1.7.9 Benefit-Cost Analysis*. The benefits should exceed the costs:

$$B/C = \frac{\Delta\ user\ benefits}{\Delta\ investment\ \text{cost} + \Delta\ maintenance - \Delta\ residual\ value}$$

$$= \frac{Present\ Value\ of\ Benefits}{Present\ Value\ of\ Costs} \geq 1$$

Step 2) Find the cost and value differences:

	Existing	Replacement	Δ
Initial Investment:	0	$85,000	**$85,000**
Salvage Value:	$12,000	$25,000	**$13,000**
Annual Costs, 1-5 years:	$10,000	$3,000	**$7,000**
Annual Costs, 6-20 years:	$14,000	$4,000	**$10,000**

Step 3) Costs include the initial investment minus the present value of the projected salvage value. Calculate the present value of the total cost using the Interest Rate Table of $i=5.00\%$ in *Chapter 1.7.10* of the Handbook:

$$Cost = \$85{,}000 - \$13{,}000(P/F,\ i=5\%,\ n=20\ years)$$
$$= \$85{,}000 - (\$13{,}000)(0.3769) = \$80{,}100.30$$

Step 4) Benefits include the cost savings between the existing and the replacement annual costs. The future annual costs from 6-20 years must be brought to year 6, then that value brought to the present. Calculate the present value of the total cost using the Interest Rate Table of $i=5.00\%$ in *Chapter 1.7.10* of the Handbook:

$$\text{Benefit} = \$7{,}000(P/A,\ i=5\%,\ n=5\ years) + \$10{,}000(P/A,\ i=5\%,\ n=15\ years)(P/F,\ i=5\%,\ n=5\ years)$$
$$=(\$7{,}000)(4.3295) + \$10{,}000(10.3797)(0.7835) = \$111{,}631.45$$

Step 5) Calculate the benefit-cost ratio:

$$B/C = \frac{\Delta\ user\ benefits}{\Delta\ investment\ \text{cost} + \Delta\ maintenance - \Delta\ residual\ value} = \frac{\$\ 111{,}631.45}{\$\ 80{,}100.30}$$
$$= 1.39\ (>1\ OK)$$

Notes:

- CERM Chapter 87
- $B/C > 1$ is expected to deliver a positive net present value and therefore you could immediately eliminate solution B.

5) Which of the following is calculated by subtracting depreciation from initial cost?

A) Capitalized costs
B) Rate-of-return
C) Inflation
D) <u>Book value</u>

Search each answer to find which is correct.

6) Which three Best Management Practices are most suitable for sediment control?

A) <u>Silt fence, straw wattle, compacted earthen berm</u>
B) Mulching, check dams, rock bags
C) Straw wattle, mulch blanket, geotextile filter fabric
D) Gradient terraces, hay bales, soil roughening

Notes:

- Understand the difference between erosion control (erosion prevention) and sediment control (preventing the movement of sediment that has already been eroded).

7) If no stratum is encountered above this depth, what is the minimum depth of exploration for embankment foundations?

A) <u>Equal to twice the embankment height</u>
B) Equal to a quarter of the embankment height
C) Between 1 and 2 times the wall height
D) 10 feet

Search *embankment foundation* and navigate to *Chapter 3.7.1 Subsurface Exploration and Planning.*

8) A direct shear test is performed on a sample of sandy soil. Horizontal stress is slowly applied to the sample until it fails in shear at 40 kPa. Vertical stress of 95 kPa is applied to the sample. The soil sample's angle of internal friction (degrees) is most nearly:

A) 23
B) 24
C) 26
D) 35

FIND: Angle of internal friction, ϕ

Step 1) Search *shear test* in the Handbook and navigate to *Chapter 3.3.1 Shear Strength - Total Stress.* Because sandy soil is cohesionless $(c=0)$, use the cohesionless soils equation from the Shear Strength figure:

$$\tau=\sigma_n \tan\phi$$

Step 2) The shear stress, τ, at failure equals 40 kPa. The normal stress, σ, equals 95 kPa. Rearrange the equation to solve for the angle of internal friction:

$$\tau=\sigma_n \tan\phi$$

$$\Rightarrow \phi=\tan^{-1}\left(\frac{40\ kPa}{95\ kPa}\right)$$

$$\phi=22.83°$$

Notes: CERM Chapter 35

9) A passenger car is traveling on a road made of portland cement concrete at a speed of 40 mph. The driver sees a stop sign and presses on the brakes in order to bring the car to a complete stop. The road is sloped downward at a 3.15% grade. The equation for stopping distance is provided. The coefficient of skidding friction, *f*, is 0.33. Most nearly, from how far away (ft) should the stop sign be visible?

A) 237
B) 294
C) <u>325</u>
D) 8,979

$$s_{stopping} = vt_p + s_b = v_{1,\ mph}t_p + \frac{v^2_{1,\ mph} - v^2_{2,\ mph}}{30(f+G)}$$

FIND: Stopping sight distance, $s_{stopping}$ (ft)

Step 1) Calculate the stopping sight distance:

$$s_{stopping} = vt_p + s_b = v_{1,\ mph}t_p + \frac{v^2_{1,\ mph} - v^2_{2,\ mph}}{30(f+G)}$$

$$= \left(40\frac{mi}{hr}\right)\left(\frac{5280\,ft}{mi}\right)\left(\frac{hr}{60\,min}\right)\left(\frac{min}{60\,sec}\right)(2.5\,sec) + \frac{\left(40\frac{mi}{hr}\right)^2 - \left(0\frac{mi}{hr}\right)^2}{30(0.33 + (-.0315))}$$

$$= 325.34\,ft$$

Notes:

- CERM Chapter 75
- t_p is the braking perception-reaction time, which is approximately 2.5 seconds for 90% of the population per AASHTO's Green Book. Therefore, 2.5 seconds is generally a fair assumption for t_p if it is not given.

10) A single-family residence is being constructed in an arid area with high seismic activity. Which construction type is most likely to be used?

A) Concrete bearing walls
B) Masonry bearing walls
C) <u>Wood framing</u>
D) Steel framing

Although this information is not necessarily in the Handbook, general basic knowledge of building construction may be helpful. Masonry building systems should not be used in areas with high seismic activity. Concrete bearing walls and steel framing systems are generally not used in residential construction, as they are relatively expensive compared to wood framing methods. Wood framing methods perform the best during seismic activity.

11) A California Bearing Ratio, CBR, test is performed to determine the suitability of a soil for use as a subbase in pavement sections. If the field CBR is determined to be 70, what type of soil is this likely to be?

A) CL
B) SM
C) GP
D) <u>GW</u>

Search *CBR* and navigate to the table in Chapter 3 of the Handbook entitled Typical CBR Values.

12) Most nearly, what is the velocity (fps) through the pipe for the system shown?

A) 12
B) <u>14</u>
C) 37
D) 48

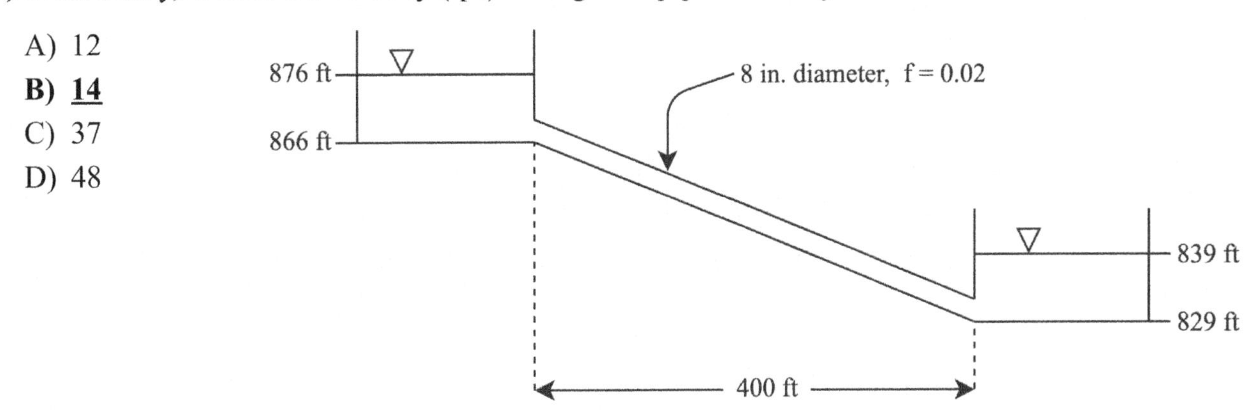

FIND: Velocity, v (fps)

Step 1) This should be automatically recognizable to you as a head loss equation. Search *head loss* in the Handbook and navigate to *Chapter 6.2.3.1 Head Loss Due to Flow*. Use the water elevations at both reservoirs as your reference points. Deduce from the diagram that there are no variations in pressure nor velocity between the two reservoirs.

$$z_1 = z_2 + h_L \Rightarrow h_L = z_1 - z_2$$

Step 2) Calculate head loss, H_L:

$$H_L = 876 \text{ ft} - 839 \text{ ft} = 37 \text{ ft}$$

Step 3) Calculate the length of pipe:

$$\sqrt{(400\ ft)^2 + (37\ ft)^2} = 401.71\ ft$$

Step 4) Calculate the velocity:

$$h_L = h_f = f\frac{Lv^2}{D2g}$$

$$\Rightarrow v = \left(\frac{h_L 2gD}{fL}\right)^{\frac{1}{2}} = \left(\frac{(37\ ft)(2)\left(32.2\frac{ft}{s^2}\right)\left(\frac{8\ in}{12\ \frac{in}{ft}}\right)}{(0.02)(401.71\ ft)}\right)^{\frac{1}{2}} = 14.06\ \frac{ft}{s}$$

Notes:

- CERM Chapters 16 and 17
- Although it doesn't affect the answer on this problem, always watch out whether you're given the actual length of pipe or the horizontal distance of pipe. That's an easy way to get tricked!

13) Which is the most correct equation to determine the ratio of total reinforcement area to the cross-sectional area of a concrete column?

A) Depth of equivalent rectangular stress block/area of shear reinforcement
B) Gross area of concrete section/total area of longitudinal reinforcement
C) Center-to-center spacing of torsional reinforcement/design flexural strength
D) Total area of longitudinal reinforcement/gross area of concrete section

Search *ratio of total reinforcement area* and navigate to *Chapter 4.3.2 Design Provisions.*

14) What is the most typical way to repair a foundation that has been settled?

A) Underpinning
B) Retaining wall construction
C) Installation of french drains
D) Soil replacement

This information is not necessarily in the Handbook. However, understanding or at least being aware of various construction and repair techniques is likely a good idea. Underpinning is the process of strengthening the foundation of an existing building or other structure.

15) Web yielding or web crippling, two types of local buckling in steel beams, can most easily be reduced or eliminated by the use of:

A) Intermediate stiffeners and flange stiffeners
B) Bearing stiffeners and exterior enamel
C) Exterior enamel and flange stiffeners
D) Wooden exterior reinforcement and double thick beams

Although this information is not necessarily in the Handbook, the incorrect options can be eliminated using a basic understanding of beams. Stiffeners are secondary plates or sections that are attached to beam webs and/or flanges to stiffen them against planal deformations. Enamel and wood reinforcement would not work well against deformations.

16) An activity-on-node diagram for a project is shown below with the activity durations in days. The float time (days) of Activity E is most nearly:

A) 12
B) <u>8</u>
C) 10
D) 6

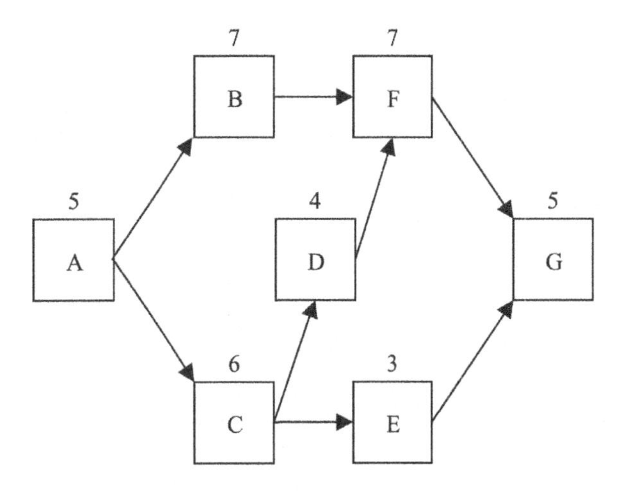

FIND: Float time of Activity E (days)

Step 1) Search *activity-on-node* in the Handbook and navigate to *Chapter 2.4.1.1 CPM Precedence Relationships*. Complete a forward pass and a backward pass on the diagram such that:

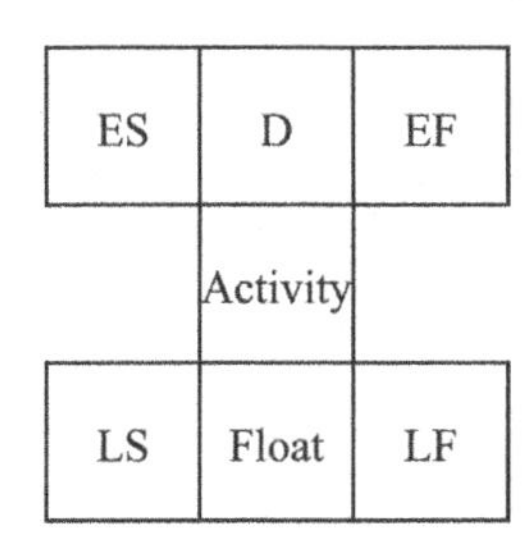

$$ES = EF_{latest\ of\ all\ predecessors}$$
$$EF = ES + Duration$$
$$LF = LS_{earliest\ of\ all\ successors}$$
$$Float = LS - ES = LF - EF$$
$$LS = LF - Duration \rightarrow LF = LS + Duration$$

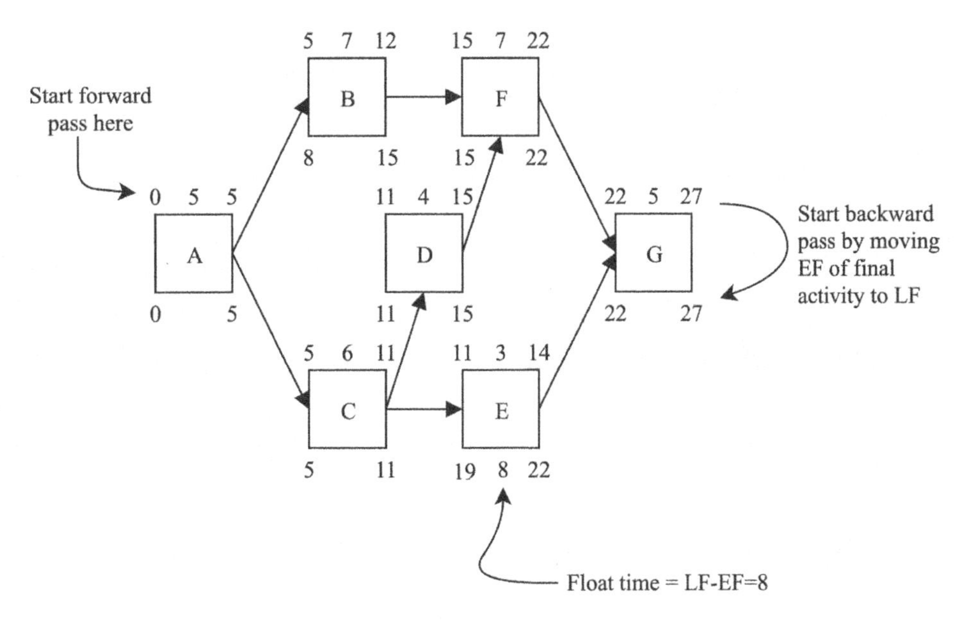

Step 2) Calculate the float time of Activity E = LF - EF = 22 - 14 = 8 days.

Note: CERM Chapter 86

17) The following measurements were taken for four ring soil samples. The moisture content (%) of the sample is most nearly:

A) 7.95
B) 8.25
C) 15.15
D) 28.00

FIND: Moisture content (%)

Total recovered mass of the four rings:	934 g
Average mass of empty ring sample container:	54 g
Average height of a single ring:	1.12 in
Dry unit weight of the soil:	58 lbs/ft^3
Average diameter of a ring sample:	3.51 in

Step 1) Calculate the total mass of the soil sample, m_{soil}, from the mass of the field sample, m_{field}, and the mass of the four rings, m_{ring}:

$$m_{soil} = m_{field} - (\#_{rings})\, m_{ring}$$
$$= 934\ g - (4\ rings)(54\ g) = 718\ g$$

Step 2) Calculate the total volume of the four rings, V_{total}:

$$V_{total} = (\#_{rings})(A_{rings})(H)$$
$$= (4\ rings)\,\pi r^2 H$$
$$= (4\ rings)(\pi)(1.76\ in)^2(1.12\ in) = 43.60\ in^3$$

Step 3) Calculate the density of the recovered soil sample, ρ_{soil}:

$$\rho_{soil} = \frac{m_{soil}}{V_{total}} = \frac{718\ g}{43.60\ in^3} = 16.47\frac{g}{in^3}$$

Step 4) Convert the soil sample density to total unit weight, γ_{total}:

$$\gamma_{total} = \left(\frac{16.47\ g}{in^3}\right)\left(\frac{lbs}{453.60\ g}\right)\left(\frac{12\ in}{ft}\right)^3 = 62.74\ \frac{lbs}{ft^3}$$

Step 5) Calculate the moisture content, w:

$$w = \frac{\gamma_{total}}{\gamma_{dry}} - 1 = \left(\frac{62.74\frac{lbs}{ft^3}}{58\frac{lbs}{ft^3}} - 1\right) \cdot 100\% = 8.18\%$$

Notes:

- CERM Chapter 35
- Although the equations aren't in the Handbook, this is a test of your ability to use your knowledge to make simple math equation from the information given

18) An underpass is being designed in a city. Given the following diagram with the elevations shown, what is most nearly the elevation (ft) at point F?

A) 454.50
B) 498.50
C) 513.50
D) 511.00

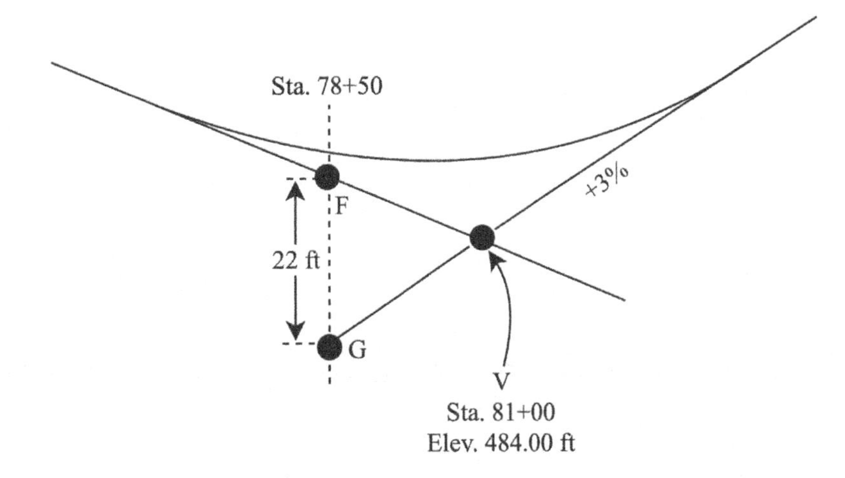

FIND: Elevation at point F (ft)

Step 1) Calculate the distance between points V and F:

$$x_{(81+00)-(78+50)} = Sta.81+00 - Sta.78+50 = 2.50\ stations\ (250\ ft)$$

Step 2) Calculate the elevation at point G:

$$elev_G = elev_V - (slope)\,x_{(81+00)-(78+50)}$$
$$= 484\,ft - (3\%)(250\,ft)$$
$$= 476.50\,ft$$

Step 3) Calculate the elevation at point F:

$$elev_F = elev_G + 22\,ft$$
$$= 476.50\,ft + 22\,ft = 498.50\,ft$$

Notes:

- For vertical curves: L, length, is the horizontal length between the PVC and PVT (*not* the length of the actual curve along the road!).
- For horizontal curves: L is the actual length of the curve and stationing is *along* the curve.

19) A hydraulic jump forms at the toe of a concrete rectangular spillway with uniform width. Before the jump, the water surface level is 8 ft above the apron with a velocity of 35 ft/s. The energy loss in the jump (ft) is most nearly:

A) 0.25
B) <u>3.25</u>
C) 20.50
D) 21.25

FIND: Energy loss (ft)

Step 1) Search *hydraulic jump* in the Handbook and navigate to *Chapter 6.4.8 Rapidly Varied Flow and Hydraulic Jump*. Calculate the depth after the jump:

$$y_2 = -\frac{1}{2}y_1 + \sqrt{\frac{2v_1^2 y_1}{g} + \frac{y_1^2}{4}}$$

$$= -\frac{1}{2}(8ft) + \sqrt{\frac{2\left(35\frac{ft}{s}\right)^2(8ft)}{32.2\frac{ft}{s^2}} + \frac{(8ft)^2}{4}}$$

$$= 21ft$$

Step 2) Calculate the energy loss (see *Chapter 6.4.8.6 Energy Loss in Horizontal Hydraulic Jump*)

$$\Delta E = \left(y_1 + \frac{v_1^2}{2g}\right) - \left(y_2 + \frac{v_2^2}{2g}\right) \approx \frac{(y_2 - y_1)^3}{4y_1 y_2}$$

$$= \frac{(21ft - 8ft)^3}{4(8ft)(21ft)} = 3.27ft$$

Notes:

- CERM Chapter 19
- Remember that the flow rate is always the same before and after a hydraulic jump!

20) Which of the following is usually not an applicable method for improving subgrade?

A) <u>Using extruded geogrid when the subgrade California Bearing Ratio (CBR) = 5</u>
B) Using woven geotextile when the subgrade CBR = 2
C) Using extruded geogrid when subgrade CBR = 1
D) Using woven geotextile when the subgrade is firm or stiff, and the total subbase thickness is 160 mm.

Navigate to the Subgrade Improvement Methods table in the Handbook.

21) Which AASHTO specification is used to measure the relative proportion of plastic fines and dust to sand size particles in material passing the No. 4 sieve?

A) <u>T176</u>
B) D2419
C) C88
D) T210

Navigate to the table in the Handbook entitled Other Tests for Aggregate Quality and Durability.

22) An investor is considering the construction of a small water treatment facility with the expected lifetime of 35 yrs. Most nearly, what is the present value of the investment ($)?

A) 14,989.84
B) <u>-15,850.40</u>
C) 16,317.00
D) -196,190.40

Initial project cost = $900,000
Annual maintenance = $12,000
Salvage value = $90,000
Annual profit = $65,000
$i = 5\%$

FIND: Present value of net profit, P ($)

Step 1) Search *economics* in the Handbook and navigate to *Chapter 1.7 Engineering Economics*. Draw a diagram to visualize:

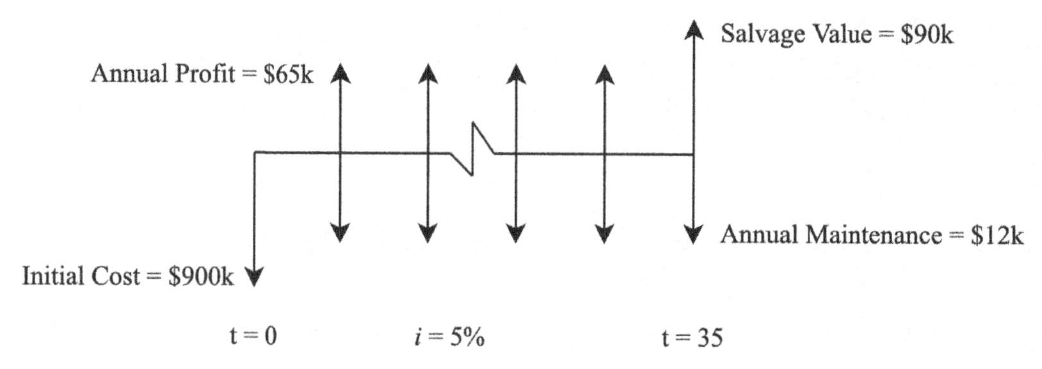

Step 2) Write out your equation based on the diagram: $P = -P_1 - P_2 + P_3 + P_4$ where:

P = Present value of investment
P_1= Initial cost
P_2 = Present cost of annual maintenance
P_3 = Present value of salvage
P_4 = Present value of annual profit

Step 3) Calculate each component using the interest rate tables in *Chapter 1.7.10*:

$P_1 = \$900,000$

$P_2 = A(P/A,\ i = 5\%,\ n = 35\ years)$
$= (\$12,000)(16.3742) = \$196,490.40$

$P_3 = F(P/F,\ i = 5\%,\ n = 35\ years)$
$= (\$90,000)(0.1813) = \$16,317.00$

$P_4 = A(P/A,\ i = 5\%,\ n = 35\ years)$
$= (\$65,000)(16.3742) = \$1,064,323.00$

Step 4) Calculate P:

$P = -P_1 - P_2 + P_3 + P_4 = -\$900,000 - \$196,490.40 + \$16,317.00 + \$1,064,323.00 = -\$15,850.40$

So, it's a bad investment because the present value is negative (*i.e.* it does not generate a profit).

Notes:

- CERM Chapter 87
- It's easy to mix up the factors so be careful to pick the right ones!
- Take a minute to make sure you are on the right interest rate table!

23) Using the profile mass diagram shown, the total cut volume (yd^3) is most nearly:

A) 0.40×10^5
B) 2.0×10^5
C) 2.6×10^5
D) 200×10^5

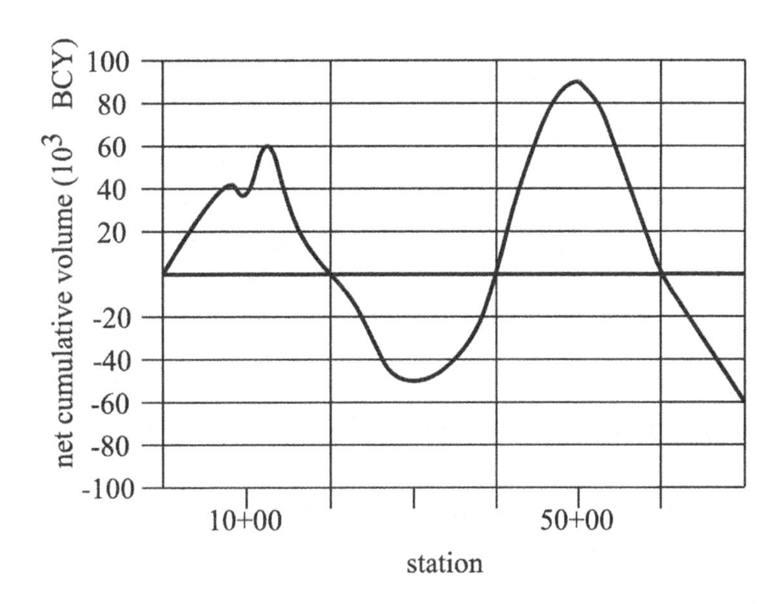

FIND: Total cut volume, V_{cut} (BCY)

Step 1) Search *construction earthwork* in the Handbook and navigate to *Chapter 2.1.4.1 Mass Diagrams and Profile Diagrams*.

Step 2) Mark the given diagram for visualization:

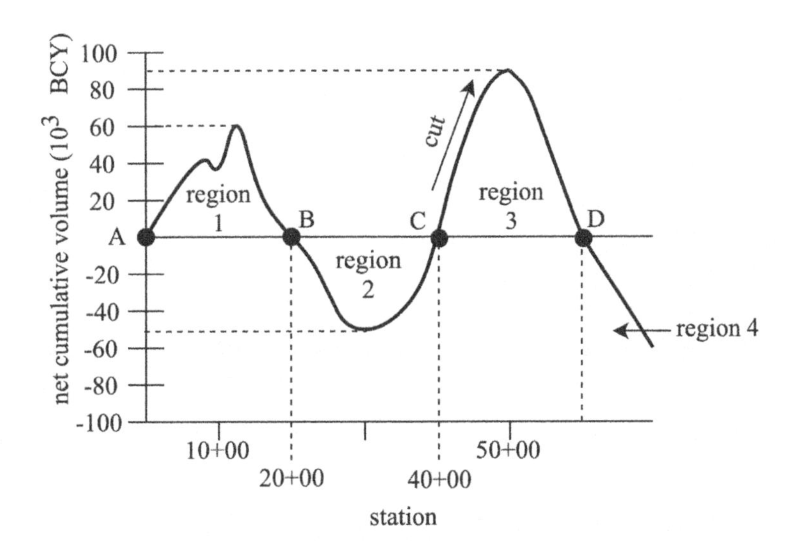

Step 4) Calculate the BCY per region:

Region 1: Mass diagram slopes up and comes back down to zero, indicating that this region requires a cut (excavation) in the beginning where the cumulative value goes up. The curve reaches a high value at approximately 60 x 10^3 BCY.

This volume of cut is then used to fill (embankment), which reduces excess soil and results in a decrease of cumulative volume. The fill volume finally breaks even at station B to bring the curve back to zero. Therefore, the volume of cut from point A (Sta. 0+00) to point B (Sta. 20+00) is $V_1 = 60 \times 10^3$ BCY.

Region 2: In this region the mass diagram begins by sloping down, indicating fill (embankment). It reaches a low point of approximately 50 x 10^3 BCY. Then cut occurs, which produces excess soil and results in an increase in cumulative volume. The volume breaks even at point C (Sta. 40+00). Therefore the volume of cut from station B to station C is $V_2 = 50 \times 10^3$ BCY.

Region 3: $V_3 = 90 \times 10^3$ BCY

Region 4: There is no cut because the curve slopes down, indicating fill (embankment) through the entire region. The problem asks for cut volume, so Region 4 isn't necessary to calculate.

Step 5) Calculate the total cut:

$$
\begin{aligned}
V_{total} &= V_1 + V_2 + V_3 \\
&= (60 \, x \, 10^3 \, BCY) + (50 \, x \, 10^3 \, BCY) + (90 \, x \, 10^3 \, BCY) \\
&= 200 \, x \, 10^3 \, BCY \, (2.0 \, x \, 10^5 \, BCY)
\end{aligned}
$$

Notes:

- CERM Chapter 80
- On mass diagrams, curves going up indicate cut (excavation) and curves going down indicate fill (embankment). Balance points are where the cut and fill lines cross the x-axis, or baseline.
- The negative values used on the mass diagram are simply used to differentiate between positive and negative net cumulative volumes. When you excavate soil from the ground, you are left with a pile of soil (positive value). When you place this soil in an embankment, you reduce the volume of available soil (negative value). Hence the term "balance point" where the net cumulative volume line crosses zero.

24) Which of the following statements is false regarding the relationship between lateral earth pressure and retaining walls?

A) Cohesive soils impose less pressure on retaining walls than cohesionless soils.
B) At-rest earth pressure imposes very nearly zero strain in the soil.
C) Passive earth pressure is the pressure in front of the wall that is a result of the wall moving towards the soil.
D) Active earth pressure is present behind a retaining wall that moves towards and compresses the remaining soil.

Search *lateral earth pressure* and navigate to *Chapter 3.1.3 Coulomb Earth Pressures* and deduce the false answer from the diagrams showing active and passive cases.

Notes: CERM Chapter 37

25) A fixed W12x96 column has a radius of gyration of 5.35 in. along the strong axis and 2.95 in. along the weak axis. The Euler buckling load of the column (kips) is most nearly:

A) 1,882
B) 1,992
C) 4,129
D) 4,264

$K = 1$
$E = 29x10^6$ psi
$A = 35.25$ in^2
$L = 18$ ft

FIND: Euler buckling load, F_e (kips)

Step 1) Search *buckling* in the Handbook and navigate to *Chapter 1.6.8 Columns* where you'll find the equation for critical buckling stress for long columns:

$$\sigma_{cr} = \frac{P_{cr}}{A} = \frac{\pi^2 E}{\left(\frac{K\ell}{r}\right)^2} \Rightarrow P_{cr} = \frac{\pi^2 EA}{\left(\frac{K\ell}{r}\right)^2}$$

Step 2) Make a quick sketch to help visualize that we are looking for r_y, the radius of gyration around the weak axis:

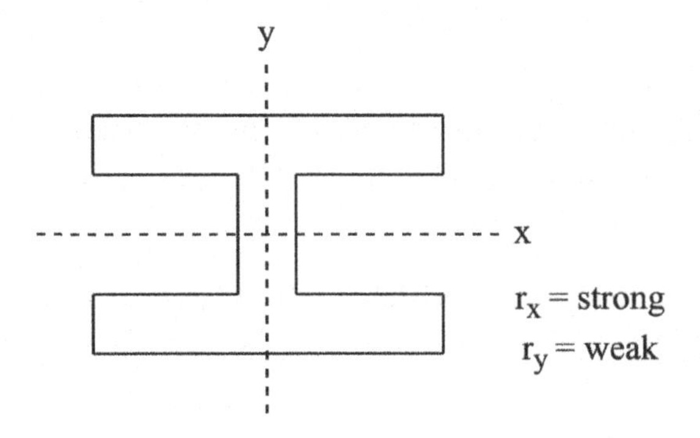

Step 3) Because the buckling will happen along the weak axis, solve the equation about r_y, or $r_{weakest}$:

$$P_{cr} = \frac{\pi^2 EA}{\left(\frac{K\ell}{r_{weakest}}\right)^2} = \frac{(\pi^2)(29x10^6\ in^2)(35.25\ in^2)}{\left(\frac{(1)(18\,ft)\left(\frac{12\ in}{ft}\right)}{2.95\ in}\right)^2}$$

$$P_{cr} = 1,881,886\ lbf = 1,882\ kips$$

Notes:

- CERM Chapter 45
- Because the K parameter is used, we know the column is fixed at both ends.
- If the problem tells you *how* the column is pinned without giving a K value, we can find the K value in *Chapter 1.6.8 Columns.*
- The designation of the wide flange beam specifies the width and weight per unit length. Therefore, W12x96 means 12 inches deep and 96 lbs/ft in this problem.
- Remember that the symbol σ means stress!

26) During peak conditions, a 10 in. water supply pipe line provides 0.67 cfs at the end of a 13,000 ft long supply line. What is the pressure head difference (ft) between the intake and discharge?

A) **125**
B) 126
C) 132
D) 133

Intake elevation = 325 ft
Discharge elevation = 196 ft
C = 200

FIND: Pressure head difference, ΔP (ft)

Step 1) Search *hazen-williams* in the Handbook and navigate to *Chapter 6.3.1.2 Circular Pipe Head Loss*. Solve for head loss due to friction, h_f (don't forget to convert inches to feet!):

$$h_f = \frac{4.73L}{C^{1.852}D^{4.87}}Q^{1.852} = \frac{4.73(13,000\,ft)}{200^{1.852}(0.83\,ft)^{4.87}}0.67\,cfs^{1.852} = 3.97\,ft$$

Step 2) Calculate the pressure difference using Bernoulli's equation, which simplifies because $v_1 = v_2$:

$$\frac{p_1}{\gamma} + \frac{v_1^2}{2g} + z_1 = \frac{p_2}{\gamma} + \frac{v_2^2}{2g} + z_2 + h_f$$

$$\Rightarrow \frac{p_1}{\gamma} + z_1 = \frac{p_2}{\gamma} + z_2 + h_f$$

$$\Rightarrow z_1 - z_2 - h_f = \frac{p_2}{\gamma} - \frac{p_1}{\gamma} = \Delta P$$

$$325\,ft - 196\,ft - 3.97\,ft = 125.03\,ft$$

Notes: CERM Chapters 17 and 19

27) A developer is looking at building a 2-story building on a site in a mild climate that does not freeze. The boring log indicates that the upper 4 ft of the soil profile is sandy peat. The sandy peat is underlain by 38 ft of dense sand, followed by 21 ft of stiff clay. What is the most suitable type and depth of foundation for the building?

A) Shallow foundation at 1 ft
B) Shallow foundation at 5 ft
C) Pile foundation at 7 ft
D) Pile foundation at 32 ft

The upper 4 ft of soil is sandy peat, which is not a good foundation soil. Below this is a dense sand underlain by stiff clay, which is good foundation soil especially for a relatively low-rise building. The most suitable choice is a shallow foundation at 5 ft, just below the bad soil.

Pile foundations, on the other hand, should be used a) when there is a layer of weak soil at the surface that cannot support the weight of the building and therefore the building loads have to bypass the weak layer, or b) when a building has a heavy, concentrated load such as in a high rise structure.

28) The foundation wall is to be 8 ft deep and 15 in. thick with a plan view shown. The volume of concrete (yd^3) required to be brought to the site for the foundation is most nearly:

A) 25
B) 64
C) 66
D) 67

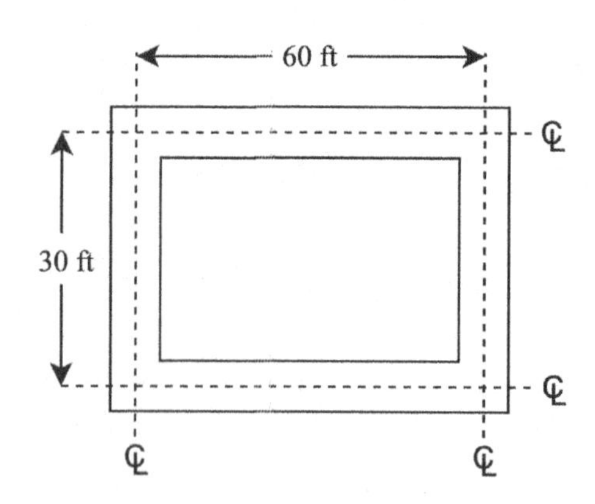

FIND: Volume of concrete, V (cy)

Step 1) Calculate the total volume of concrete and convert to cubic yards:

$$\left(60\,ft + 15\,in\left(\frac{ft}{12\,in}\right)\right)(2)(8\,ft)\left(15\,in\left(\frac{ft}{12\,in}\right)\right) +$$
$$\left(30\,ft - 15\,in\left(\frac{ft}{12\,in}\right)\right)(2)(8\,ft)\left(15\,in\left(\frac{ft}{12\,in}\right)\right) = 1,800\,ft^3\left(\frac{yd}{3\,ft}\right)^3 = 66.67\,yd^3$$

Notes:

- On quantity take off problems, take precise note of what the measurement is (i.e. inside vs centerline vs outside of wall) before you start calculating!

29) The watershed unit hydrograph of a 1-hr storm has a peak discharge of 225 cfs. If a 1-hr storm drops 1.75 in. net precipitation, the peak discharge (cfs) is most nearly:

A) 11
B) 129
C) 130
D) 394

FIND: Peak discharge, Q (cfs)

Step 1) The unit hydrograph is developed by dividing every point on the overland flow hydrograph by the average excess precipitation, $P_{ave,\ excess}$. Therefore, the hydrograph for a storm can be found by multiplying the ordinates of the unit hydrograph (including the peak) by the excess precipitation:

$$Q_p = (Q_{unit\ hydrograph})(P_{ave,\ excess}) = \left(225 \frac{ft^3}{s \cdot in}\right)(1.75\ in) = 393.75\ \frac{ft^3}{s}$$

Notes:

- Excess precipitation is the volume of rainfall available for direct surface runoff. It is equal to the total amount of rainfall minus all abstractions including infiltration, interception, and depression storage. Because runoff is the excess precipitation after abstractions, "excess" and "total" are generally interchangeable.

30) In analyzing a determinate statics problem, which configuration and orientation of forces describes one in which all the forces act at the same point?

A) Collinear force system
B) Singular point force system
C) Coplanar force system
D) Concurrent force system

This can be found in *Chapter 1.5.12 Concurrent Forces* of the Handbook.

Notes: CERM Chapter 41

31) A permeability test is conducted with a sample of soil that is 123 mm long and has a diameter of 64 mm. The flow is 1.45 mL in 484 seconds and the head is kept constant at 201 mm. The coefficient of permeability (m/yr) is most nearly:

A) <u>18.0</u>
B) 18.5
C) 19.1
D) 19.2

FIND: Coefficient of permeability (m/yr)

Step 1) Search *permeability* in the Handbook and navigate to *Chapter 3.8.5.1 Laboratory Permeability Tests*:

$$K = \frac{QL}{tAh} = \frac{QL}{t\pi r^2 h}$$

$$= \frac{(1.45\ mL)(123\ mm)}{(484\ sec)(\pi)(32\ mm)^2(201\ mm)} x \frac{\left(\frac{1000\ mm^3}{mL}\right)\left(\frac{365\ day}{year}\right)\left(\frac{24\ hr}{day}\right)\left(\frac{60\ min}{hr}\right)\left(\frac{60\ sec}{min}\right)}{\frac{1000\ mm}{m}}$$

$$= 17.98\ \frac{m}{yr}$$

Notes:

- CERM Chapter 35
- The term Q stands for Volume in this equation in the Handbook.
- The coefficient of permeability often has the units of ft/s or m/s.

32) Although the footing area of a foundation is obtained from unfactored service loads, the footing thickness and reinforcement are calculated from factored loads. The width of a reinforced concrete wall footing is 5 feet and the factored wall load per unit length is 25 kips. What is the factored load per unit length (kips/ft) at failure?

A) <u>5</u>
B) 25
C) 15
D) 50

FIND: Factored load per unit length, q_u (kips/ft)

Step 1) Use simple math to determine the factored load per unit length at failure.

$$q_{factored} = \frac{P_{factored}}{B} = \frac{25\ kips}{5\ ft} = 5\frac{kips}{ft}$$

Notes: CERM Chapter 55

33) According to OSHA, what is the noise dose (%) of 90 dBa during a standard 8-hr work day? Does it require abatement?

A) **8.5%; No abatement required**
B) 9.0%; Abatement required
C) 40%; No abatement required
D) 1600%; Abatement required

Find: Noise dose (%)

Step 1) Search *noise dose* and navigate to *Chapter 2.6.2 Work Zone and Public Safety* in the Handbook for the equation for noise dose, *D:*

$$D = 100\, x \sum \frac{C_i}{T_i}$$

Step 2) Deduce from the table that at a noise level of 120 dBa, the permissible time of exposure is 0.125 hours. Calculate *D*:

$$D = 100\, x \sum \frac{C_i}{T_i} = 100\, x \,\frac{8\ hrs}{95\ hrs} = 8.42\%$$

Step 3) Noise abatement is required when D > 100%, therefore noise abatement is not required under these circumstances.

34) For determining slope stability for cohesive foundation soils, which of the following is the most likely reason for consolidating some samples to a higher than existing in situ stress?

A) To determine the increase of deflection needed to reduce the pore pressure to the required levels.
B) To determine the results of the field vane shear test.
C) To determine the shear strength in relation to depth.
D) **To determine strength gain of clay due to consolidation under staged fill heights.**

Search *slope stability* in the Handbook and navigate to *Chapter 3.6.6 Slope Stability Guidelines*. Deduce the correct answer from the table entitled Slope Stability Guidelines for Design.

35) Two contiguous 4-ac watersheds shown below are served by an adjacent 500 ft long storm drain. Inlets are placed along the storm drain to collect runoff from the respective watershed. The storm drains flow full. The intensity of a storm is 5 in./hr. The flow time (min.) from inlet 1 to inlet 2 is most nearly:

A) 0.29
B) 4.78
C) 11.30
D) 41.80

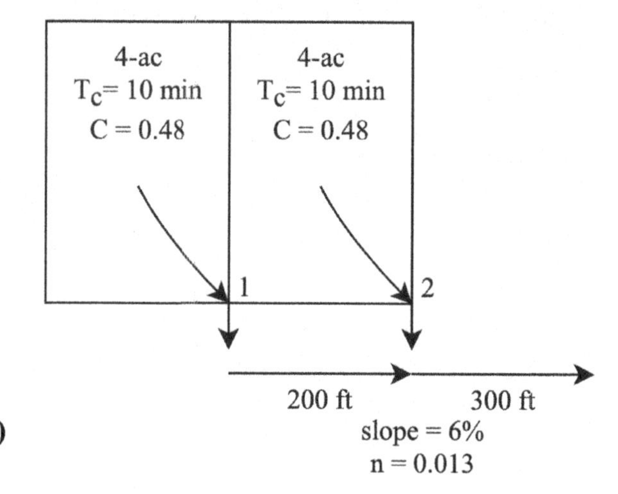

FIND: Flow time between inlets 1 and 2, t (minutes)

Step 1) Search *rational method* and navigate to *Chapter 6.5.2.1 Rational Formula Method.* Calculate peak discharge for the first watershed:

$$Q = CIA = (0.48)\left(5\frac{in}{hr}\right)(4\ ac) = 9.6\ \frac{ft^3}{s}$$

Step 2) Calculate the pipe diameter, *D*, using the equation for minimum diameter of the circular pipe under full flow conditions in *Chapter 6.4.5.3 Flow in Circular Pipe* of the Handbook:

$$D = \left(\frac{C_0 Qn}{\sqrt{S}}\right)^{\frac{3}{8}} = \left(\frac{(2.16)\left(9.6\ \frac{ft^3}{s}\right)(0.013)}{\sqrt{0.06}}\right)^{\frac{3}{8}} = 1.04\ ft$$

Step 3) Calculate the velocity in the pipe between inlets 1 to 2:

$$v_{full} = \frac{Q}{A} = \frac{9.6\ \frac{ft^3}{sec}}{\pi\left(\frac{1.04\ ft}{2}\right)^2} = 11.30\frac{ft}{sec}$$

Step 4) Calculate the time it takes for the runoff to flow from inlet 1 to 2:

$$t_{1-2} = \frac{L}{v} = \frac{200\ ft}{\left(11.30\frac{ft}{s}\right)\left(60\frac{s}{min}\right)} = 0.29\ min$$

Notes:

- CERM Chapters 19 and 20
- Remember that Q = CIA has inconsistent units.

36) If the water table is at the bottom of Layer 1 as shown below, the total settlement (in.) of the normally consolidated clay layer for the soil profile shown below is most nearly:

A) 0.16
B) 0.23
C) 1.88
D) 2.76

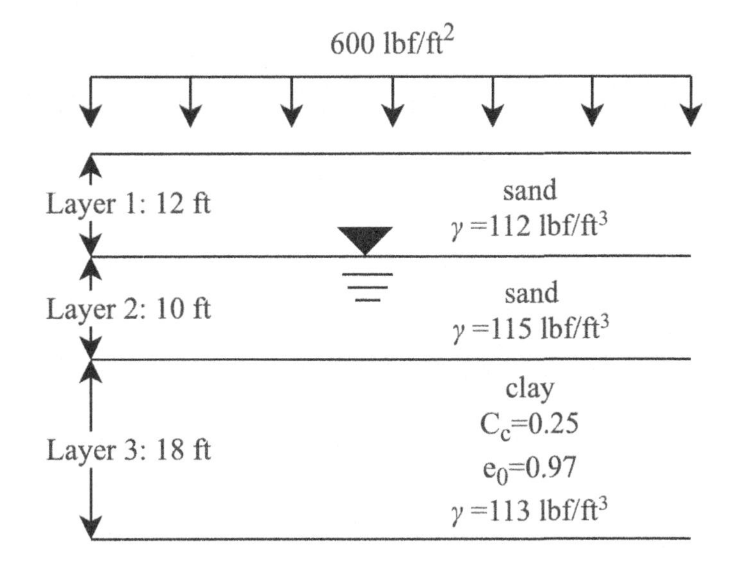

FIND: Settlement, S (inches)

Step 1) Search *settlement* in the Handbook and navigate to *Chapter 3.2.1 Normally Consolidated Soils*. Find the appropriate equation (amended to account for the water table) to solve for settlement, S:

$$S_C = \sum_1^n \frac{C_c}{1+e_o} H_o \log_{10}\left(\frac{p_f}{p_o}\right) = \sum_1^n \frac{C_c}{1+e_o} H_o \log_{10}\left(\frac{p'_o + \Delta p}{p'_o}\right)$$

Step 2) Find the total original stress, p_o, at the middle of the clay layer:

$$p_o = \left(112\frac{lbf}{ft^3}\right)(12\,ft) + \left(115\frac{lbf}{ft^3}\right)(10\,ft) + \left(113\frac{lbf}{ft^3}\right)\left(\frac{18}{2}ft\right) = 3,511\frac{lbf}{ft^2}$$

Step 3) Calculate the pore pressure at the middle of the clay layer:

$$\mu = \gamma_{water} h = \left(62.4\frac{lbf}{ft^3}\right)\left(10\,ft + \frac{18}{2}ft\right) = 1,185.60\frac{lbf}{ft^2}$$

Step 4) Calculate the original effective stress, p'_o, at the middle of the clay layer:

$$p'_o = p_o - \mu = 3,511\frac{lbf}{ft^2} - 1,185.60\frac{lbf}{ft^2} = 2,325.40\frac{lbf}{ft^2}$$

Step 5) Calculate the total settlement, *S*, of the normally consolidated clay layer:

$$S_C = \sum_1^n \frac{C_c}{1+e_o} H_o \log_{10}\left(\frac{p'_o + \Delta p}{p'_o}\right) = \frac{0.25}{1+0.97} \cdot 18\,ft \cdot \log_{10}\left(\frac{2,325.40\frac{lbf}{ft^2} + 600\frac{lbf}{ft^2}}{2,325.40\frac{lbf}{ft^2}}\right) = 0.23\,ft$$

Step 6) Convert to inches:

(0.23 ft)(12 in/ft)= 2.76 in

Notes:

- CERM Chapter 40
- Remember in consolidation calculations, the *original effective pressure*, p'_o, is calculated at the middle of the clay layer because it represents the average conditions of the clay. This is however not the case for settlement of the whole layer.
- Read through the "where" section after the equations for more information about each parameter.
- The *original* versus *vertical* pressure can be confusing. The original and vertical pressures are essentially the same, however the *increase* in vertical pressure (Δp) is from the additional applied pressure shown on the diagram.

37) Which of the following gel times (min.) would you use for a low concentration silicate grout?

A) 200
B) 320
C) 440
D) 880

Search *grout* in the Handbook and navigate to *Chapter 3.12.2 Grouting*. Find the correct gel time in the table entitled Physical Properties of Chemical Grouts.

38) The 25 ft long Oak beam has a 2-in. x 8-in. cross section. The beam is subject to a uniform load of 0.7 kip/ft and a point load of 6.25 kips at 6 ft from Point A, as shown. Most nearly, where is the maximum moment measured from point A (ft)?

A) 4.36
B) 6.02
C) <u>10.36</u>
D) 13.50

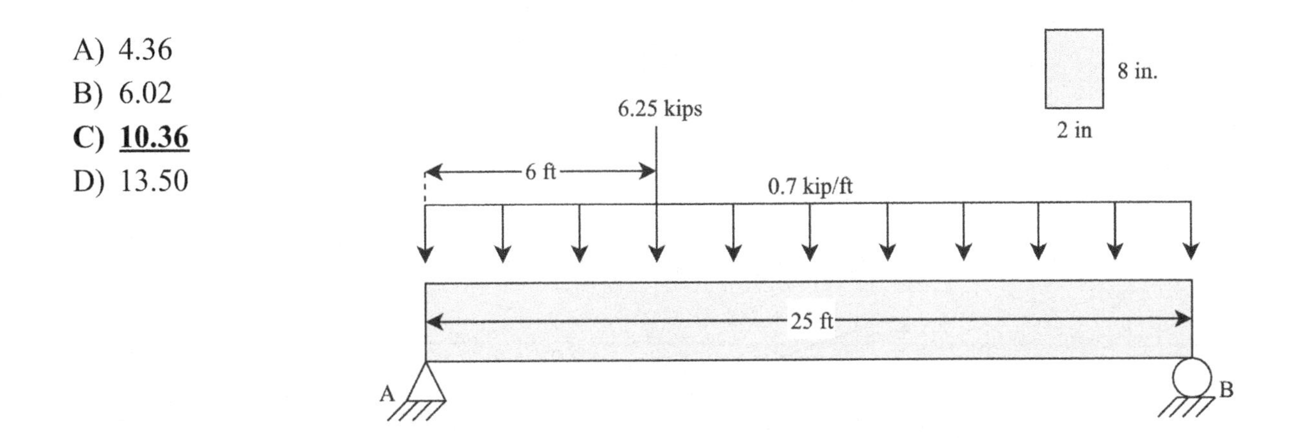

FIND: Distance from point A to max. Moment, x (ft)

Step 1) Search *shear moment* in the Handbook and navigate to the Shear, Moments, and Deflections table in *Chapter 4.1.7*. The first scenario (Simple Beam - Uniformly Distributed Load) applies.

Step 2) Simplify the beam into its components:

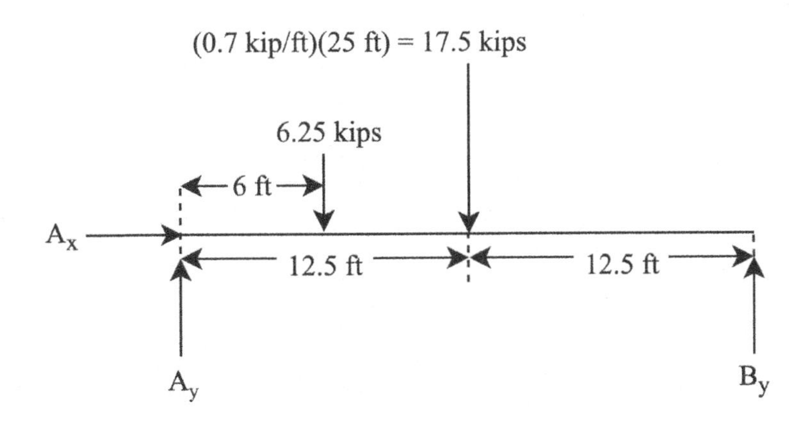

Step 3) Solve for the vertical reaction force at support B, B_y:

$$M_A = \Sigma Fd = 0;\ (17.5\ kips)\ (12.5\ ft)\ +\ (6.25\ kips)(6\ ft)\ = B_y(25\ ft)$$

$$B_y = 10.25\ kips$$

Step 4) Find the vertical reaction force at A, A_y, by taking the sum of forces in the x- and y-directions:

$$F_x = 0;\ A_x = 0$$

$$F_y = 0;\ A_y = 6.25\ kips + 17.5\ kips - 10.25\ kips$$

$$A_y = 13.50\ kips$$

Step 5) Draw the shear diagram to find the maximum moment, M_{max}:

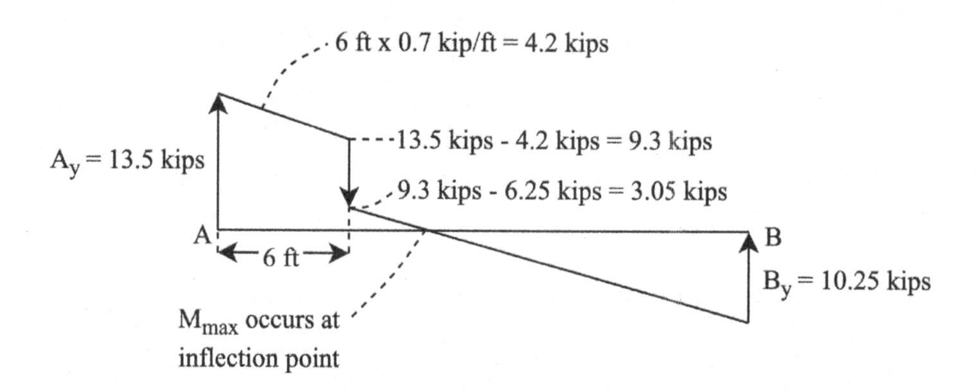

Step 6) Looking at the latter half of the shear diagram, find M_{max} from the shear diagram above using the slope or like triangles:

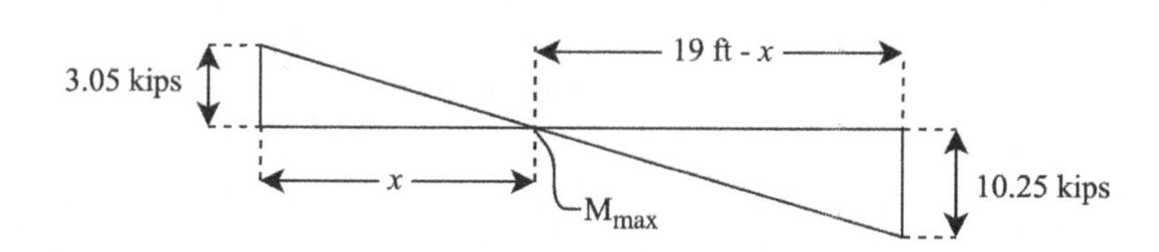

Slope:

$$Slope = \frac{\Delta y}{\Delta x} \Rightarrow 0.7\frac{kips}{ft} = \frac{3.05\frac{kips}{ft}}{x} \Rightarrow x = 4.36\frac{kips}{ft}$$

Like triangles:

$$\frac{x}{3.05\ kips} = \frac{19\,ft - x}{10.25\ kips}$$

$$x = 4.36\,ft$$

Step 7) Calculate the distance from support A to M_{max}: 6 ft + 4.36 ft = 10.36 ft

Notes: CERM Chapter 44

39) A public circular storm sewer is designed to carry a peak flow of 8 ft^3/s. The municipality requires that the depth at peak flow is a maximum of 75% of the pipe diameter. Based on topography, the required sewer slope is 2.5%. The required sewer diameter (in.) is most nearly:

A) <u>14</u>
B) 21
C) 18
D) 12

n = 0.012

Table: Depth of flow (d), Diameter (D), Area (A), Wetted Perimeter, and Hydraulic Radius (r_h) of a Partially Filled Circular Pipe

d/D	*area/D^2*	*Wetted perimeter/D*	*r_h/D*
0.73	0.6143	2.0488	0.2995
0.74	0.6231	2.0714	0.3006
0.75	0.6318	2.0944	0.3017

FIND: Sewer diameter (inches)

Step 1) Search *open channel flow* in the Handbook and navigate to *Chapter 6.4.5.1 Manning's Equation.*

$$Q=\left(\frac{1.486}{n}\right)AR_H^{\frac{2}{3}}S^{\frac{1}{2}}$$

Step 2) Discern from the given table that where $d/D_{full}=0.75$:

$$\frac{A}{D^2}=0.6318 \qquad \text{and} \qquad \frac{r_h}{D}=0.3017$$

$$\Rightarrow A=0.6318D^2 \qquad \Rightarrow r_h=0.3017D$$

Step 3) Insert A and r_h calculated above, and calculate the diameter:

$$Q=\left(\frac{1.486}{n}\right)AR_H^{\frac{2}{3}}S^{\frac{1}{2}}$$

$$8\frac{ft^3}{s}=\left(\frac{1.486}{0.012}\right)\left(0.6318D^2\right)(0.3017D)^{\frac{2}{3}}(0.025)^{\frac{1}{2}}$$

$$8\frac{ft^3}{s}=5.56464D^{8/3}$$

$$D=1.15ft\ (14\ in)$$

Notes:

- CERM Chapter 19
- Learn how to use the equation solver on your calculator to solve tricky equations faster!

40) The equation for beam flanges in compression is given. A compact steel beam has the parameters shown. Most nearly, what is the required allowable stress yield, F_y, (ksi) for the flanges to be in flexural compression?

A) 36
B) 50
C) <u>15</u>
D) 18

$$\frac{b_f}{2t_f} \le 0.38\sqrt{\frac{E}{F_y}}$$

Flange width, b_f = 24 inches
Flange thickness, t_f = ¾ inch
E = 29,000 ksi

FIND: Allowable stress yield, F_y (ksi)

Step 1) Balance the given equation:

$$\frac{b_f}{2t_f} \le 0.38\sqrt{\frac{E}{F_y}}$$

$$\frac{24\ in}{2(0.75\ in)} \le 0.38\sqrt{\frac{29,000\ ksi}{F_y}}$$

$$\frac{24\ in}{2(0.75\ in)} \le 0.38\sqrt{\frac{29,000\ ksi}{15\ ksi}}$$

Notes:

- Instead of spending time balancing the equation, speed things up by substituting the given answers until it's balanced.

<u>- END OF SOLUTIONS VERSION 3 -</u>

Made in the USA
Coppell, TX
19 February 2022